W0231882

Reinhard Kurth Walther K. Schwerdtfeger (Eds.)

Current Topics in Biomedical Research

With 53 Figures and 15 Tables

Springer-Verlag Berlin Heidelberg GmbH

Professor Dr. Reinhard Kurth
Priv.-Doz. Dr. Walter K. Schwerdtfeger

Paul-Ehrlich-Institut
Bundesamt für Sera und Impfstoffe
Paul-Ehrlich-Straße 51–59
D-6070 Langen

ISBN 978-3-642-77081-4 ISBN 978-3-642-77079-1 (eBook)
DOI 10.1007/978-3-642-77079-1

Library of Congress Cataloging-in-Publication Data.
Current topics in biomedical research / Reinhard Kurth, Walter K. Schwerdtfeger (eds.).
Includes bibliographical references and index.

1. Immunology--Congresses. 2. Virology--Congresses. I. Kurth, Reinhard, 1942- . II. Schwerdtfeger, W. K. (Walter K.), 1949- . [DNLM: 1. Immunity--congresses. 2. Research--congresses. QW 504 C976] QR180.3.C87 1992 616.07'9--dc20 DNLM/DLC for Library of Congress

Originally published by Springer-Verlag Berlin Heidelberg New York in 1992
Softcover reprint of the hardcover 1st edition 1992

25/3145-5 4 3 2 1 0 – Printed on acid-free paper

Preface

This volume contains the papers presented at the symposium "Concepts in Biomedical Research", which was held on May 3-5, 1990, on the occasion of the Opening Conference of the new facilities of the Paul Ehrlich Institute at Langen, near Frankfurt.

The Paul Ehrlich Institute in its capacity as the Federal Agency for Sera and Vaccines, is responsible for the control and release of immunobiological drugs in Germany. In addition, its activities embrace a wide variety of biomedical projects in bacteriology, virology, immunology, allergology and veterinary infectious diseases.

Current scientific challenges accounting for a significant proportion of the work of the Paul Ehrlich Institute, particularly the urgent need to prevent the further spread of the acquired immunodeficiency syndrome, were reflected in the presentations given by the invited speakers, whose contributions have been updated during the compilation of this volume. The authors not only present recent results but also give overviews of their respective areas of research and discuss future research priorities. Thus the articles are of interest both to specialists in infectious diseases and to all those interested in progress at the frontiers of present-day biomedical research.

The editors wish to thank the authors for their contributions to this volume and for their helpful cooperation, as well as the publisher, Springer-Verlag, for producing the book.

Langen, October 1991

Reinhard Kurth
Walter K. Schwerdtfeger

Contents

The Authors

The following summarizes the principal academic and professional achievements of the authors. Within the limited space available, it is impossible to list all the honors, awards, memberships and positions held by the scientists who contributed to this volume.

Renato Dulbecco Prof., M.D., Dres. h.c.; President, Salk Institute for Biological Studies, La Jolla, California; Member, National Academy of Sciences (Washington); Foreign Member, Royal Society (London); Albert Lasker Basic Medical Research Award 1964; Nobel Prize for Physiology or Medicine 1975.

Manfred Eigen Prof., Dr. rer. nat., Dres. h.c.; Director, Department of Biochemical Kinetics, Max Planck Institute for Biophysical Chemistry, Göttingen; Foreign Honorary Member, American Academy of Arts and Sciences (Boston); Foreign Associate, National Academy of Sciences (Washington); Foreign Member, Royal Society (London); Nobel Prize for Chemistry 1967; Paul Ehrlich and Ludwig Darmstädter Prize 1992.

D. Carleton Gajdusek Prof., M.D., Dres. h.c.; Director, Laboratory of Slow, Latent and Temperate Virus Infections, and Chief, Central Nervous System Studies Laboratory, National Institute of Neurological Disorders and Stroke, National Institutes of Health, Bethesda; Member, National Academy of Sciences (Washington); Member, American Academy of Arts and Sciences (Boston); Nobel Prize for Physiology or Medicine 1976.

Dieter Haustein Prof., Dr. rer. nat.; Head, Department of Allergology/Immunopharmacology, Federal Agency for Sera and Vaccines (Paul Ehrlich Institute), Langen.

Maurice R. Hilleman Prof., Ph.D., Dres. h.c.; Director, Merck Institute for Therapeutic Research, West Point, Pennsylvania; Professor of Pediatrics, University of Pennsylvania; Member, Expert Advisory Panel of the World Health Organization; Member, National Academy of Sciences (Washington); Member, American Academy of Arts and Sciences (Boston); Albert Lasker Public Service Award 1983.

Joachim R. Kalden Prof., Dr. med.; Head, Medical Clinic III, University of Erlangen; Editor, *Rheumatology International*; Chairman, WHO/IUIS Standardization Committee; Member, Senate Commission for Cancer Research of the Deutsche Forschungsgemeinschaft.

Reinhard Kurth Prof., Dr. med.; President, Federal Agency for Sera and Vaccines (Paul Ehrlich Institute), Langen; Member, Advisory Panel on Biological Standardization of the World Health Organization; Member, Scientific Council of the Paul Ehrlich Foundation.

Jonas Salk Prof., M.D., Dres. h.c.; Founding Director, Salk Institute for Biological Studies, San Diego; Member, Expert Advisory Panel on Virus Diseases of the World Health Organization; Fellow, American Academy of Arts and Sciences (Boston); Senior Member, Institute of Medicine, National Academy of Sciences (Washington); developer of the poliomyelitis vaccine; Albert Lasker Clinical Research Award 1956; Robert Koch Medal 1963.

Michael Schlesinger Prof., Dr. med.; Chairman, Paul Ehrlich Center for the Study of Leukemic and Normal White Blood Cells and Hubert H. Humphrey Center for Experimental Medicine and Cancer Research, Hebrew University, Jerusalem.

Hans-Gerhard Schwick Prof., Dr. phil., Dr. med. h.c.; Chairman, Board of Directors, Behringwerke AG, Marburg; Member, Advisory Board of the Society for Immunology; Member, Scientific Council of the Robert Koch Foundation; Robert Koch Medal 1973.

Fritz Stern Prof., Ph.D., Dr. h.c.; Professor of History, Columbia University, New York; publicist; Member, Board of Trustees of the German Marshall Fund; Fellow, American Academy of Arts and Sciences (Boston); Lionel Trilling Book Award 1977.

Howard M. Temin Prof., Ph.D., Dres. h.c.; Professor of Biological Sciences, University of Wisconsin, Madison; Executive Editor, *Molecular Carcinogenesis*; Member, National Academy of Sciences (Washington); Foreign Member, Royal Society (London); Fellow, American Academy of Arts and Sciences (Boston); Albert Lasker Basic Medical Research Award 1974; Nobel Prize for Physiology or Medicine 1975.

Otto Westphal Prof., Dr. rer. nat., Dr. med. h.c.; Retired Director, Max Planck Institute for Immunobiology, Freiburg; Co-Founder, *European Journal of Immunology*; Life Member, Collegium Allergologicum Internationale; Foreign Honorary Member, American Academy of Arts and Sciences (Boston); Founding President, Gesellschaft für Immunologie; Paul Ehrlich and Ludwig Darmstädter Prize 1968; Robert Koch Medal 1983.

Paul Ehrlich in His Time: A Historian's View

F. Stern

Columbia University, New York

As a historian, unversed in natural sciences, I will try to put a few aspects of science to the historical framework of the past 150 years: its role in history has been steadily changing, and its many different aspects have never been fully recognized by historians. There is a famous, but shameful gap between the humanities and the natural sciences, which were described by C.P. Snow as the two cultures. When I was taking an interest in some representatives of science in Germany in the decades preceding and following the First World War - Haber, Einstein and their circle, as well as the politicoscientific culture surrounding them - I was interested in man, in man with his love for science, which was still innocent at that time. The passion leading to great achievements was to be felt, but what were the conditions of involvement and success? What did "science as a profession" mean? What was the scientific ethos of that time, the prerequisite for creativity?

How can the historian understand and explain the numerous sources of scientific progress? He can consider an individual representative, for instance, Paul Ehrlich, who was fully concentrated on his work, dedicated to research, practically isolated from the world, and restless, yet at the same time influenced by his environment. He was a research scientist in his own time, in a time when researchers had a special significance.

Helmholtz once spoke of the "inner psychological history" of science, a tempting subject. Even Einstein emphasized the "interaction of scientific achievement and character." I am fascinated by the human power, by the perseverance required during the time of construction. There were research geniuses. The word "genius" has a distinct meaning; in German there is even a demonic association. A genius, whether an artist or a scientist, is particularly vulnerable, sometimes insecure, and always endangered. It was in Ehrlich's time that Thomas Mann began to analyze the vulnerability of the artist, and he continued these analyses as long as he lived. But isn't there a certain congeniality of artists and great scientists? There are many differences, too. An artist is in most cases alone and steadily endangered, whereas most scientists have a home, enjoy friendly relations, even friendship among colleagues, which can even soothe occasional unkindness on the part of colleagues.

In Paul Ehrlich's life, as before him in the lives of Virchow, Pasteur, and Koch, the same vocabulary appears - genius, leader, general - words to describe these personalities, as the public recognized them. And it was a historical moment when the heroic individual research worker created a new world, sometimes even by risking his own life.

The first original age of geniuses was the beginning of the *Sturm und Drang* period, a time when great poets and philosophers created a new literature, a new art, in which the names Mozart and Beethoven implied "genius", a time when a new image of man was created, a new language and a new feeling for the language, and when there was hope for a new type of education. It was also the time of great political and moral reformers in Prussia and Southern Germany, a time, when Napoleon was the embodiment of personality and its development. It was a time when man was measured by talent and achievement, and when men's lives were not determined by birth or rank. It was an unusual time, in which modernism took its rise - the practical, political life in America, the intellectual life in Europe. It was a time when people did not avoid speaking of genius and when, in Germany in particular, the dark and mysterious powers of genius were recognized.

Paul Ehrlich was born in a second age of geniuses, which has not yet been fully recognized. It was a time of great inventions, a time when a scientific and experimental approach was taken to medicine, and when the first great progress was made. There was a widespread and unconditional belief in empiric or positivistic sciences and that man was able to understand and control nature. An example: Rudolf Virchow, physician, research worker, and fighter for liberalism against Bismarck, said in 1865 during the 40th assembly of German scientists and physicians: "Science has become our religion." Or in 1873: "We, too, have a belief, the belief in progress by realizing the truth." This belief dominated the thinking of most scientists until the end of the First World War. In January 1911, when the *Kaiser-Wilhelm-Gesellschaft* was founded, Emil Fischer said that there was no future in conquering colonial empires, but "it is chemistry and all other natural sciences which offer unrestricted possibilities." It was a belief which was supported by evidence; it was the time of triumph of practical findings and battling against disease, epidemics, and death.

This belief was also the basis of scientific ethos. Wilhelm Oswald, also a winner of the Nobel prize, said in 1906 about the suicide of Ludwig Boltzmann that a true scientist had to suffer from pain and loneliness. "This is one of the laws to which nearly all servants of the severe goddess Science are subject; their lives end in sorrow, the more complete their devotion, the greater their sorrow." Einstein, too, spoke of a "temple of science" and referring to Max Planck's untiring perseverance and patience he said that the emotional state which generates these achievements is similar to religious feelings or falling in love. Daily endeavors are not based on intention or programs, but on direct needs. These were public confessions. Many unpublished letters by scientists such as Einstein, Haber, and Willstätter show modesty and dissatisfaction, emotions which can also generate a belief in holy things. The goddess is so supreme that the modesty of the individual can be understood: the modesty of Paul Ehrlich is well known, and Haber remarked in a letter to Willstätter: "I am pleased to know about things which are beyond my ability, and I am happy to admire them."

Natural sciences, however, also require public appreciation and support. In the last 20 years of the nineteenth century, there was a kind of international rivalry among scientists, even though an international cooperation of research workers was taken absolutely for granted. There were no boundaries in science, but the large states in Europe realized that science was also an economic factor, not only a fight for prestige or scientific Olympic Games, that there was power involved, economic as well as military power. There is much proof that statesmen and political leaders, all of whom had a classical education, recognized the new role of research, in particular of basic research. In this context I would like to mention the establishment of the Institute Pasteur in 1889. In his last speech in 1892, shortly before his death, Pasteur, an ardent French patriot, spoke of his "invincible belief that science will triumph over ignorance and war, the nations will unite, not to destruct, but to construct, and the future will belong to those who were engaged in helping the suffering mankind."

In Germany at the time of Emperor William there was a growing engagement with science; its utility was recognized. William II, who was fascinated by modern technology, was a strong advocate of science, which for him had a nationalistic pathos. The continuous progress in Germany especially in the fields of medicine and chemistry, also evident by the great number of German Nobel prize winners, was at first based on spontaneous cooperation among industry, science, the universities and the state. At the end of the nineteenth century a complex with scientific, economic, and governmental features developed in Germany, supported by patrons, which enabled Germany to gain a unique position. This kind of organized cooperation was exemplary, even though somewhat dangerous, and it is now being established in the United States. There was also much objection to the rise of science from men who suspected danger in the belief in science, e.g., Nietzsche and Max Weber, and there were numerous groups who felt that their moral monopoly in education or in religion was threatened by positivistic science. Nietzsche, convinced that Darwinism was dangerous, was skeptical about the alleged advantages of science for life. Max Weber recognized the human and philosophical limits of science. Although he advocated unbiased science, he said in his speech on science as a profession: "Nothing is dear to man, unless he does it with passion." Science, however, does not answer the all-important questions: "What are we going to do? How are we to live?" For Weber, science was a profession in the sense of protestant ethics, but there is no salvation in the disenchanted world.

There were groups who felt threatened by the rise and claims of the sciences, threatened as far as their own position was concerned: The humanists, the professors in the predominant fields of that time, history, philology, and philosophy, looked at chemists or scientists with disdain. They believed them to be upstarts, handicapped by positivism and materialism, who even entered into relations with industry and business. This seemed to be an offense against German idealism, which reminds me of Nietzsche who said: "Not to recognize himself is the wisdom of the idealist. The idealist is a being who has his reasons for not knowing anything about himself; he even is wise enough to leave these reasons in the dark." It is known and understandable that the church objected to science. As in many intellectual conflicts, there were misunderstandings and fear involved; many of the great scientists themselves were religious and hoped to help mankind without destroying belief in the mysterious.

It was a world of progress into which Paul Ehrlich was born and to which he contributed considerably. Born in Strehlen, near Breslau, in 1854, into a wealthy Silesian-Jewish family, his interest in science was encouraged by his cousin Carl Weigert. He obtained his doctorate, but during his medical training he became fascinated by physiological research, and he came to recognize the immense possibilities of industrially manufactured stains for physiological tests. In 1878, Friedrich Frerich, one of the most renowned German internists and one of Bismarck's physicians, took young Ehrlich on to the Charité, one of the most famous hospitals in Germany. Under Frerich's guidance Ehrlich was able to devote himself entirely to his research. Frerich recognized the genius in Ehrlich. Seriousness and hard work, concentration and forgetfulness were typical of the young Paul Ehrlich. In 1882 Robert Koch gave his famous lecture on tuberculosis. Later, Ehrlich wrote: "Everyone who took part in this lecture, was deeply moved. I have always remembered this evening clearly as my greatest scientific experience." Frerich and Koch were Ehrlich's ideals; at the same time they supported and promoted him.

When Koch died, Ehrlich wrote an obituary which had a certain significance for his own life: "Obituaries seem to be like autobiographies. It was the pioneering work of the young physician Robert Koch who proved the specificity of the various bacteria and their responsibility for infectious diseases." Ehrlich described the circumstances under which Koch did his scientific research, his dealing with problems which had also been tackled by prominent scientists and which he was able to solve with acumen and untiring work. "Genius and energy inherent in his personality released his pioneering ideas." But did not the same apply to Paul Ehrlich? "It seems a matter of fact that this man is being celebrated, a hero of science, a victorious leader in his battle against dangerous epidemics... All of us who have known him admire his research genius, his intellect, his inexhaustible efficiency, his energy, and also his courage which made him defy all dangers and enabled him to become a protector of public safety, a victorious leader in the fight against man's most dangerous enemies." This is what he wrote about his hero. Future generations had similar thoughts about Ehrlich.

In 1885 Frerich died; his successors failed to understand Ehrlich and his research, and there were hard times ahead for him and his wife Hedwig Pinkus, professionally as well as physically. Ehrlich developed tuberculosis, probably becoming infected in the Charité laboratory. He went to Egypt to regain his health, and then resumed his research activities. A few years later he was given support by Friedrich Althoff, the omnipotent ruler of Prussian universities, who had the ambition to make his universities the best worldwide. Robert Koch took Ehrlich on at the newly established Institute for Infectious Diseases. Later, owing to Althoff's and the mayor of Frankfurt's intervention, when the Royal Institute for Experimental Therapy was founded in Frankfurt, its head was Paul Ehrlich.

His achievements in various fields are well known. His publications demonstrate his energy: his first publication appeared when he was 23 years old; in 1914 there were 212 publications which built up his international reputation.

For him, nature was friend and enemy alike. He developed drugs which rescued mankind from some diseases and epidemics. Before the diphtheria antitoxin was developed by Ehrlich and Behring, 50% of all children died from this disease. Ehrlich's most famous success was Salvarsan, a drug which cured syphilis, a disease which claimed a growing number of victims and which was also considered a vice. A few decades before, a famous English doctor praised syphilis as God's punishment for sinners. Paul Ehrlich freed men of this disease, a fact which was criticized by some hypocrites of that time.

His colleagues admired not only his success, but also his working methods. There are some remarks which greatly impress the historian. Richard Willstätter, for example, writes about Ehrlich's "love of stains", an expression which characterized the passionate scientist. More important, however, is Willstätter's general judgment: "We are reminded of great thinkers and artists of old times when we admire the universality of Ehrlich's work and when we recognize his versatility as a chemist, although he was deeply involved in organic chemistry. As a chemist, Paul Ehrlich is nobody's disciple; he has never been influenced by any current of his time; wide and new fields were opened up by his intuition and personal power."

His old friend Albert Neisser, a well-known dermatologist, wrote: "Ehrlich's lifework was not based on coincidental discoveries, he owes his successes to hard work. Ehrlich ist not a 'discoverer' in the usual sense of the word, someone who stumbles over gold when ploughing his fields, or who is suddenly inspired. His achievement can be compared to that of a marksman who knows his weapon, his bullets and their path, and whose eyes are directed to the target which he wants to hit. The first bullets might go astray, but the number of hits grows until the experienced marksman succeeds in hitting the mark. His principle: 'We must learn how to aim' applies not only to his therapy, but to his whole research work."

He worked internationally. Like Robert Koch, he travelled to the remotest parts of the world to study epidemics on the spot. He gave lectures all over Europe and in America; the international medical congress held in London in 1913 celebrated him as the world's most famous research scientist. One of his closest colleagues was Japanese. As an individual he anticipated and improvised the international role which modern research institutes now try to organize.

Let me add a final historical thought. Paul Ehrlich was and remained a Jew; at the same time, however, he regarded himself as a German, and abroad he was celebrated as a German. In Germany he suffered from discrimination, and disappointment - the bitter fate of all Jews. I do not believe that any of the great Jewish scientists was able to escape from the prejudice. Neither christening nor extraordinary achievement could change the social standing. In spite of his success, there were often moments in Ehrlich's life when he was faced with discrimination and hostility. And he was troubled about the obstacles which his nongenius follow-Jews had to overcome. The injuries he suffered were relieved by his colleagues and superiors who highly appreciated him and offered him their support - Frerich, Koch, Althoff and many others. Although on a higher level there seemed to be a harmonious creativity in Germany, the prejudice was almost universally present - in this context I do not want to use the word antisemitism because of its racist and populistic tinge.

The resentment towards Jews had a certain fascination. Certain qualities were regarded as typically Jewish, i.e., ambition, succeeding in one's intentions, the allegedly unscrupulous aspiration to be successful and to make money. To be able to have someone else to blame strengthens one's moral superiority, although it might also have concealed a double anxiety - the anxiety of infection, that one might have the same qualities, and also the fear that the Jews might contest their Christian colleagues' positions, their intellectual patrimony.

German Jews have experienced animosity and friendship. The obstacles created by the prejudice often had the opposite effect: antisemitism in general was a stimulus for Jews to work extremely hard to safeguard their interests in sometimes unconscious competition. Antipathy towards Jews whose social assimilation was advanced is also seen in the lives of Ehrlich and his colleagues. The Imperial Society benefited from their prejudices, a certain psychic satisfaction and increased readiness to make sacrifices on behalf of Jewish scientists. Germany and the German industry took great advantage of Ehrlich's and Haber's achievements, and they contributed to Germany's worldwide preeminence. A sociological study was done on this phenomenon in medicine and the natural sciences: the gifted young scientist was discriminated in the beginning, got minor positions, was passed over, and had to remain a lecturer for a long period of time, so that he was able to devote his time to his own specialist research work. Paul Ehrlich did not live to see the decline of this remarkable Jewish-Christian cooperation or to experience the vileness of later years. Although he was given recognition and awards, he never received a full professorship, but he was appointed honorary professor, received an honorary doctorate, won the Nobel prize and was appointed Privy Councillor with the title of Excellency, a rare award in Germany.

Since his time, the situation of science has changed considerably. Value, and costs, of the natural sciences are widely recognized today. Paul Ehrlich used to say that he could work in a barn. Modern institutes are not barns at all and prove that things have changed in science. This change also means increased dangers, highly specialized work, dissipation. Nearly 90 years ago, Max Weber warned that the word could become the truth: "Specialists without intellects, Epicureans without heart: This cipher flatters itself on having climbed a stage of humanity never reached before." The natural sciences lost their innocence and the progress of technology gives rise to anxiety and threatens life. The white coat has no longer the same meaning as before. Today, there is a great demand for responsibility. Science has become more of a "big science" and is less dominated by geniuses such as Koch, Ehrlich and, lately, Robert Oppenheimer.

The memory of Paul Ehrlich is a memory of a great age and a reminder at the same time. The German catastrophe struck science, but was partly generated by science. For German science, the name Paul Ehrlich is also a great moral obligation. Science has been offered a unique chance in history, a second chance of a new beginning in a new Europe. May it take advantage of this chance, comply with the moral and political engagement required today and achieve what Paul Ehrlich used to wish: money, patience, skill, and luck.

The Dawn of Immunology: The Era of Paul Ehrlich

O. Westphal

Chemin de Ballalaz 18, CH-1820 Montreux

Merging of Medicine and Natural Sciences

In an eventful period in which German medicine was just turning away from Romanticism and Natural Philosophy and when the scientific way of thinking and experimentation was making strong progress, Paul Ehrlich was born on 14th March 1854 in Strehlen, near Breslau in Silesia. Only one day later, Emil Behring was born in Hausdorf, East Prussia. A few years before Ehrlich's birth, a severe typhoid epidemic had struck the region of Upper Silesia, and the Prussian government sent the young pathologist Rudolf Virchow (1821-1902) from Berlin to the province to carry out an inspection; this resulted in his famous propositions for governmental help [1]. A few years later, the same Virchow, then Professor in Würzburg, developed his fundamental ideas on cellular pathology, in which he postulated that all manifestations of disease could be reduced to disturbances in living cells. He also expressed his now well-known principle that "All cells originate from other cells".

In 1856, when Paul Ehrlich was 2 years old, William Henry Perkin (1839-1907), a young student at the Royal College of Chemistry under the directorship of August Wilhelm Hofmann (1818-1892; Fig. 1), synthesized by chance the first aniline dye, the purple colored mauvein. Hofmann and Perkin became pioneers in the field of the chemistry and technology of synthetic dyestuffs, such as in the synthesis of the rosaniline series, which would become so important for Ehrlich.

When Paul Ehrlich was 10 years old, Louis Pasteur (1822-1895) in Paris delivered the final blow to the theory of the spontaneous generation of life [2]. As a direct consequence, 1 year later Joseph Lister (1827-1912) in Glasgow carried out the first surgical operation under antiseptic conditions [3,4]. The basis of modern bacteriology and hygiene which was thereby established was soon to be greatly extended by Robert Koch (1843-1910) [5].

The New Dyes in Histology and Biochemistry

Carl Weigert (1845-1904), a cousin of Paul Ehrlich who was older than him by 9 years, first systematically tested many of the new aniline and other dyes for the *selective* staining of healthy and diseased tissue. Weigert was an assistant at the Institute of Pathological Anatomy at Breslau University from 1868 to 1878. As early as 1870, he had demonstrated bacteria in the tissue of infected organisms using dyestuff-staining techniques. Weigert also invented a microtome which enabled serial cuts to be made from tissue preparations.

This highly creative and highly educated scientist [6] invited his young cousin Paul Ehrlich, who was then still a schoolboy in Breslau, to spend his free time in the holidays in his laboratory; thus, Weigert's influence became crucial, as Ehrlich's many visits awoke in him a "grande passion" for dyestuff histology. As a medical student, after having spent one semester in Breslau in 1872, Ehrlich moved to Strasbourg, which at that time had only a few hundred students. The histological staining experiments which he performed independently during the anatomy courses were favorably looked upon and encouraged by his teacher, Professor Wilhelm Waldeyer (1836-1921) [7].

Shortly thereafter, while studying in Freiburg, Ehrlich (Fig. 2) discovered a new cell type among white blood cells. On account of their specifically colorable granules, which he interpreted as cellular food reserves, he named them mast cells. With the further development of these histological dye studies, Ehrlich became the founder of modern blood cell teaching. It was the selectivity of the staining process that fascinated and guided him in his subsequent work. As a 23-year-old student, Ehrlich published his first work on *Aniline Dyes and their Use in*

Fig. 1. August Wilhelm Hofmann, 28 years old. **Fig. 2.** Ehrlich as a student around 1873. **Fig. 3.** Paul and Hedwig Ehrlich on holiday in Upper Silesia (1883). **Fig. 4.** Robert Koch in 1882

1
AUG. WILH. HOFMANN
IM 28. LEBENSJAHRE
2
3
4

Microscopical Technique [8]. He went to discover eosinophils by means of eosin staining, and neutrophils by staining with neutral red. Moreover, he was able to distinguish between granulocytes and lymphocytes.

After his thesis and final examinations in Leipzig, in 1876 Ehrlich became an assistant at the Medical Hospital of the Charité in Berlin under Professor Theodor von Frerichs (1819-1885). Frerichs, an excellent physician and imaginative teacher, was convinced that a thorough scientific education is indispensable for the clinical investigator. Ehrlich was given a small laboratory as a kind of scientific play corner, where he could organize the time he spent on experiments and clinical services to his own sense of responsibility. For Ehrlich it was the happiest and most concentrated time in his career. In 1883, aged 29, he married Hedwig Pinkus who contributed much to his creativity, good humor, and contentment (Fig. 3).

In the meantime, Robert Koch (Fig. 4) had been appointed director of the Public Health Administration in Berlin where he started his work on tuberculosis. Six months later, he had already claimed the discovery of the tubercle bacillus. Paul Ehrlich was present at Koch's memorable speech in 1882 which was attended by many eminent colleagues and also by Virchow, the critical pathologist (Fig. 5). Immediately after Koch's presentation, Ehrlich developed a modified staining procedure to detect tubercle bacilli in the sputum more clearly than before [9]. The method was based on the acid fastness of the bacilli, and Koch appreciated and accepted it gladly.

Ehrlich continued his blood cell studies in the Charité and became interested in leukemias and the various forms of anemias. The question of the lifespan and regeneration of red blood cells as carriers of oxygen led him to study cell respiration. Using dyestuffs *in vivo*, Ehrlich studied oxygen binding and the consumption of various cell types and their energetics in relation to the cell function. He developed indicator dyes that exist in a strongly colored oxidized form and a colorless reduced form; both forms constitute a reversible redox equilibrium. These dyes can be used to study the dynamics of cellular respiration. Thus, Ehrlich was the first to use vital coloring, and he confirmed his basic conception of the selective reactivity of certain dyes in vivo. For example, he introduced alizarin blue and especially indophenol, a blue water-insoluble dye which under reducing conditions forms an alkalisoluble colorless leuco-product [10].

At the age of 31, Ehrlich summarized his work on redox dyes and their use in vital staining in his famous book *On the Requirement of the Organism for Oxygen* [11], which appeared in 1885 and was dedicated to his teacher Frerichs. He concluded that, in the cells of the organism, oxygen-binding groups of different binding energy and thus of different oxygen saturation do exist, and he defined three such groups. Ehrlich's conclusions contrasted with the teachings of the famous physiologist Eduard Pflüger (1829-1910), then professor in Bonn, who assumed full oxygen saturation of all protoplasm under physiological conditions. Pflüger was the most formidable critic of his time. He used to say: "Criticism is the most important impulse of progress. For that reason I practise it". The young Ehrlich was enthusiastic enough to declare Pflüger to be wrong!

Ehrlich's methylene blue reaction of living nerve substance is another classic example of organotropic vital coloring [12]. In 1884, in his laboratory in the Charité, this reaction made visible not only the fine ramifications of the nerve fibers, but also the higher metabolic activity of the nerve endings.

Ehrlich was convinced that dyestuffs and tissues react with each other because of their complementary chemical structures. He took great interest in the rapidly developing field of organic chemistry. He often said he had a "plastic sense" for molecular structures. Towards the end of Ehrlich's student years in 1874, the Dutch physicochemist Henricus Van't Hoff (1852-1911), who was the same age as Ehrlich published his fundamental ideas on *The Positions of Atoms in Space* [13, 14]. Van't Hoff and, independently, Achille Le Bel (1847-1930) both proposed the tetrahedral model of the carbon atom, which formed the basis of stereochemistry and all subsequent organic chemistry in theory and practice [15, 16].

The unexpected death of Frerichs in 1886 was a great loss [17]. Frerichs successor had no comprehension of Ehrlich's scientific work and reduced research freedom in favor of clinical routine work, to the point at which the situation became intolerable for Ehrlich. As the symptoms of tuberculosis became apparent, Ehrlich and his young wife left Berlin and went to Egypt to recover. Professor Felix Pinkus who was both a relative of Ehrlich's wife and a colleague, later wrote [18]:

"His disease was diagnosed as lung tuberculosis and indeed he exhibited the clinical symptoms of this, but his main suffering was from constraints".

The Birth of Immunological Science

Restored to health, Ehrlich returned to Berlin in 1888, the year of the three emperors. In 1889, he accepted a position at the Institute of Hygiene under Robert Koch, which led him into the new field of immunology. Emil Behring (1854-1917) together with Shibasaburo Kitasato (1852-1931) had just discovered the antitoxins of diphtheria and tetanus at the same institute [19]. By coincidence, bacterial exotoxins had recently been described by Emil Roux (1853-1933) and Alexandre Yersin (1863-1943) at the Pasteur Institute in Paris. Behring's discovery of humoral immunity, exactly one century ago [19] opened up a new line of thinking: the possibility of a causative therapy of bacterial infections. This was the basis of serum therapy. In 1892, passive immunization with Behring's serum was successfully applied for the first time in treating children suffering from diphtheria.

Strongly impressed by these discoveries and stimulated by Robert Koch, Ehrlich set up model experiments with the proteinaceous plant poisons ricin and abrin, and he was able to show that antiricin and antiabrin also protected animals against the effects of these poisons [20, 21]. He also demonstrated a high immunizing efficiency via the oral route of toxin administration. In terms of the high specificity, he said:

> "Antiricin and antiabrin have no serological relation to each other. It follows that plant toxins which are otherwise so similar must necessarily also be chemically different".

We are at the roots of immunochemistry, an expression which was introduced 10 years later by the Swedish physicochemist Svante Arrhenius (1859-1927) [22]. Ehrlich and Arrhenius later met many times and had vigorous discussions about the nature of toxin-antitoxin reactions [23], which Arrhenius declared to be reversible equilibrium reactions.

In Ehrlich's first experiments, there was only little protection, but he improved this by slowly increasing the dose of the poison. Thus, a maximum of antitoxin was eventually produced that was sufficient, in animal experiments, to provide

Fig. 5. Rudolf Virchow around 1890. **Fig. 6.** Arches under the Berlin city railway. Photograph after 1900. **Fig. 7.** Friedrich Althoff around 1890. **Fig. 8.** Emil Fischer in the late 1890s

Rudolf Virchow
1882–1902

Dr. Althoff

Emil Fischer · 1902

protection against a more than 1 000-fold lethal dose of the plant toxin. Ehrlich's strategy could now be applied to diphtheria toxin. To this end Behring and Ehrlich signed a handwritten contract in 1892.

Stabsarzt Professor Dr. Behring
B e r l i n S.O.
Treptower Chaussee 8,1
"Villa Emma"

The two undersigned agree to perform experiments together with aim of obtaining diphtheria sera from mammals under the following conditions:
1. The experimental animals will be stabled under the railway construction N 278 (Fig. 6).
2. The maximum number of animals to be accommodated by Ehrlich is being set at 10 goats and 1 cow.
3. For Behring, the minimal number of goats is set at 7.
4. The animals belonging to the undersigned are to be procured by each one and are to be taken as each one's property.
5. Ehrlich pays one quarter of the total costs of maintenance (all care and observation) which are based on the current year from Oct. 15, 1893 to Oct. 15, 1894 and amount to about 7200 marks; thus Ehrlich pays 1800 marks.
6. The remedies obtained from these animals in this way will be submitted to the Hoechst dye company. The resulting sales will be shared 25% by Hoechst, and 37.5% by each of the undersigned.
7. The contract is valid until April 1897.

Berlin, 14th October 1892
Ehrlich Behring

In the years to come, both Ehrlich and Behring could indeed be satisfied with the continuous and significant reduction of diphtheria mortality.

Generation and Action of Antitoxins - The Side Chain Theory

Standardization of the antitoxic sera was essential for their production and clinical use. In 1896, Dr. Friedrich Althoff (1839-1908; Fig. 7) from the Prussian Ministry for Cultural Affairs, a brilliant man and a most influential personality in the science policy of his time, created an Institute for Serum Research and Serum Testing in Berlin-Steglitz and appointed Ehrlich director.

Work started immediately. Ehrlich now first showed that the specific toxin-antitoxin reaction is really of a chemical nature. "All observations", he said, "suggest that the reaction between toxin and antitoxin takes place in accordance with the proportions of a simple equivalence, especially since it is possible to titrate the antibodies with great accuracy - in favourable circumstances, the error with the present methods is 1%. A molecule of toxin combines with a definite and unalterable quantity of antibody" [24]. Ehrlich and colleagues from the Pasteur Institute showed (a) that antitoxin is not derived from toxin (as many had suggested), (b) that antitoxin only blocks toxicity, without chemically destroying the toxin, and (c) that antitoxic antibodies are not antibacterial. Finally, he said [24]:

"It must be assumed that the ability to combine with antitoxin is attributable to the presence in the toxin complex of a specific group of atoms with a maximum specific affinity to another group of atoms in the antitoxin complex, the first fitting the second easily, as a key does a lock, to quote Emil Fischer's well-known simile."

Fischer's respective lecture (1852-1919; Fig. 8) [25] had been delivered a few years previously in Berlin, almost arround the corner as it were and was heard attentively by Ehrlich (see also [26, 27]). The two great scientists developed mutual admiration.

In his studies, Ehrlich observed that some toxin solutions turned out to be unstable. Some weakly toxic preparations, however, still retained their full antitoxic immunizing potency, and these were termed "toxoids". It seemed that different areas in the toxin molecule were responsible for immunization and for toxicity.

In 1876, the chemist Otto Nikolaus Witt introduced a theory to explain the structural basis of the color of dyestuff molecules in terms of chromophoric groups [28]. In his publication, Witt already used the term "side chains", which, when attached to an aromatic nucleus, would confer the colored character. Ehrlich, referring to Witt, now introduced the term haptophoric groups for the immune-specific regions (today the determinants or epitopes), and toxophoric groups for the toxic regions of the toxin molecule.

Ehrlich (Fig. 9) now saw many analogies: dyestuffs react selectively with specific cell structures; oxygen binds to the respiring cell with the aid of structures with special affinity for oxygen. He thought it possible that the effect of drugs could be similarly explained. In every case, binding is the first step. A substance is toxic for a cell because of its toxophoric group, but will become effective only after its binding via affine receptors. If a substance has immunogenic properties, the haptophoric attachment to the receptive cell is the signal that triggers the immune process. But how are the antibodies in antisera being produced? Ehrlich argued [29]:

"To attribute an inventive activity to the body or to its cells (in response to toxin action), enabling them to produce new groups of atoms ... would be a return to concepts current in the past days of natural philosophy. Our knowledge of cell function and especially of synthetic processes would lead us rather to assume that in the formation of antibodies we are dealing with the enhancement of a normal cell function, and not with the creation at need of new groups of atoms. Physiological analogues of the group of specifically combining antibodies must exist beforehand in the organism or in its cells."

Ehrlich termed those cell structures which mediate binding "side chains" [30]. This led him to his famous and much discussed theory about the formation and nature of antitoxins, first published on two pages in 1897. Bacterial (or plant) toxins bind with their haptophoric group to the cell receptors. At adequate toxin concentration, the cell reacts according to a pathobiological principle expressed by Carl Weigert (Fig. 10), not only with regeneration but with overproduction of the receptors [31], which are then released into the blood stream. Circulating toxin will react with the excreted side chains, and this results in the competitive inhibition of its toxic effects. In Ehrlich's own words [32]:

"With regard to certain molecular groups of the living protoplasm which mediate the binding of toxins and are, thus, the cause of the susceptibility (sensitivity) of the organism toward a certain toxin, I have designated them "receptors". After the introduction of toxins, according to my theory of the generation of antitoxins, the receptors will be produced in excess and, finally, excreted into the blood stream as "useless" ballast. The freely circulating receptors are the antitoxins".

In a review in 1898, Behring wrote about Ehrlich's sidechain theory [33]:

"Just as the same substance in the living body, fixed in a cell, is the supposition and condition of intoxication, so it will be the reason of healing if it is dissolved in the blood fluid."

Elie Metschnikoff (1845-1916; Fig. 11) at the Pasteur Institute in Paris wrote in his widely distributed book *Immunity and Infectious Diseases*, published in 1902 [34]:

"As we know, Ehrlich does not say by what cells the side chains are being produced. In our opinion, the phagocytes fulfill this task".

However, it is fair to say that in 1899 Ehrlich and Julius Morgenroth [35] had quoted Richard Pfeiffer who, when referring to the side chain theory and antibacterial immunity the previous year, had shown experimentally that anticholera immune bodies are mainly produced in the spleen, bone marrow, and lymphatic glands. This was more generally confirmed by Ehrlich's coworker August Wassermann. The so-called reticuloendothelial system was defined much later in 1924 by the pathologist Ludwig Aschoff (1866-1942), who was 10 years younger than Ehrlich. In 1902, Aschoff wrote an excellent review about the side chain theory [36].

Moving to Frankfurt

Shortly before the end of the eventful century, in 1899, the Institute for Experimental Therapy, which was much bigger than Ehrlich's small institute in Berlin, was founded in Frankfurt and Ehrlich appointed as director. The foundation of the new institute was again under the auspices of Althoff's initiative and the active participation of Frankfurt's lord mayor Adickes. It was also strongly supported by Emil Behring (Fig. 12), who was then in Marburg, Carl Weigert, who was then director of the Department of Anatomy at the Senckenberg Institute in Frankfurt, and not least by the local dye company Hoechst. At the official opening on the 6th November 1899 in the presence of many distinguished representatives of political, cultural, and industrial life, Ehrlich gave an important talk entitled: *Mode of Action and Mechanism of Production of Antitoxins* [37].

At the end of the nineteenth century, Ehrlich (Fig. 13) could be regarded as the most creative and competent scientist in the field of humoral immunity; Metschnikoff, on the other hand, had contributed the most to the theory of cellular immunity. Thus, it appeared quite justified, after Behring had become a Nobel laureate in 1901, that in 1908 the Nobel prize for pioneering work in the area of immunology was presented to both Ehrlich and Metschnikoff.

In the Institute for Experimental Therapy, Ehrlich now continued and extended his studies on the phenomenon of immunological cell lysis, particularly hemolysis and bacteriolysis, with Morgenroth and Hans Sachs. To designate the heat-sensitive serum component that mediates immunocytolysis, Ehrlich finally introduced the term "complement".

With the side chain theory, another aspect arose for Ehrlich, which immunologists today like to paraphrase by the terms "self" and "nonself". The question arose of what happens if the organism perceives its "own" structures as "foreign" with the induction of immune reactions against "self". Ehrlich theoretically spoke of "horror autotoxicus", and he found himself confronted with such ideas again when he entered the field of cancer research.

We should mention here two aspects of Ehrlich's manner of experimental and intellectual working. One of his characteristics was the simplicity of his equipment which consisted, of course, of a good microscope, but otherwise only tubes and chemicals (Fig. 14). Henry Dale, who worked with him in 1903, later described how Ehrlich tested a sample of dye for its purity by pouring a solution of it down a suspended sheet of blotting paper and looking for the appearance of accessory adsorption bands. "Paper chromatography in embryo, it would seem", wrote Dale. Another characteristic was Ehrlich's excellent and at the same time very economical memory. Sometimes he liked to call himself a "monomaniac" who had no real interest in anything but his science. He was also amusical. But he added [38]:

"We medical scientists are confronted with an unlimited flood of literature (Fig. 15). The one, however, who is being caught by a dominating idea, is unconsciously acquiring a kind of collector and classifier enabling him to pick up only what is useful and in direct connection with his very subject. This is very convenient because, unconsciously, everything of relevance is coming readily to him and, by reflexion, will be put rightly together."

Beginning of Cancer Research

In 1903, when Ehrlich was almost 50 years of age, the cloud of an increasing incidence of cancer hung over the country, underlined by the tragic death of emperor Friedrich and later also his wife. Several wealthy people, strongly supported by Althoff, raised funds and urged Ehrlich to initiate basic research into the nature

Fig. 9. Ehrlich, 42 years old. **Fig. 10.** Carl Weigert around 1890. **Fig. 11.** Elie Metschnikoff around 1903. **Fig. 12.** Emil von Behring in 1900

9
10
Paul Ehrlich 42 Jahre
Carl Weigert
11
12
Elie Metchnikoff
nil von Behring · 1901

and treatment of cancer. Before 1903, almost no cancer research had existed, despite the accumulation of an enormous amount of phenomenological clinical data. Human tumor material was used only in pathology for histological studies, and a colleague of Ehrlich, Dr. Anton Stricker, had failed in cumbersome experiments to transplant any human tumor into smaller experimental animals. However, in 1903 the 40-year-old Danish veterinary medical doctor, Carl Oluf Jensen (1864-1934) published his results of experimental work with mouse and rat tumors and their successful transplantation. His studies showed that cancer cells can, thus, survive over many generations [39].

After Ehrlich had established collegial contact to Jensen, the first transplantable mouse tumor arrived in Frankfurt; this marked the beginning of extensive studies by Ehrlich, Hugo Apolant, and Georg Schoene. Over the course of the next few years, hundreds of spontaneous tumors from mice and rats were collected from almost all over the world, and their transplantability and general biology was screened.

Finally, by means of subcultivation, certain tumor lines, particularly from mice, became highly malignant and were found to be 100% transferable. The well-known Ehrlich ascites tumor, for example, has since been sent out to many cancer laboratories and transferred in thousands from mouse to mouse. Many other transplantable tumors were described, and this formed the basis of experimental cancer research.

At the beginning of the twentieth century, there was a continuous and highly interesting theoretical discussion on the etiology of cancer. Many scientists, but neither Ehrlich nor Weigert [40], favored a parasitic basis, which could not in any case be verified at that time.

In further studies, Ehrlich and his colleagues showed that by "immunization" (as they called it) with tumor material the animals treated in this way could acquire substantial protection against the specific tumor. However, at that time it was impossible to differentiate between true tumor immunity and transplantation immunity. Inbred strains were still not in use. In a critical review and with reference also to the first systematic studies by Leo Loeb (1869-1959) at the University of Pennsylvania on the fate of transplanted normal and tumorous tissue of rats [41], Georg Schoene pointed out that such tumor immunity could also be induced by injecting nonmalignant, normal, and particularly embryonic tissue. On the other hand, no antisera against tumors could be produced.

Consequently, Ehrlich raised the question of natural tumor immunity. He said [42]:

"Since we have now shown that cancer immunization can be achieved also with normal organ material, this is proof that pure cellular immunity exists, and I like to believe that this is also the basis of mechanisms of natural protection in man".

Following Pasteur's ideas of infectious immunity as nutritional exhaustion of the pathogen, Ehrlich found that experimental tumors could be "starved" by placing an additional tumor, thus depriving them - as he thought - of important nutrients (though of unknown nature). He called the effect "atreptic immunity" [42] (starvation immunity), and he emphasized the possible importance of atrepsia for an equilibrium between primary tumors and their metastases.

In 1909, Ehrlich summarized the results of his cancer research by saying:

"When one succeeds in protecting small animals in a simple and reliable way against infection with tumor material of the greatest virulence, then the possibility should exist of achieving the same in human beings"[42].

The great gap between experimental cancer research in animals and clinical cancer treatment of humans could not, however, be bridged. In addition, it must be remembered that tissue and cell culture outside the body had only just been introduced by the 35-year-old Alexis Carrel (1873-1944) at the Rockefeller Institute in 1908; for this he received the Nobel prize for medicine in 1912. In the most widely distributed handbook on the teaching of cancer published between 1911 and 1914, prominent clinicians expressed their view thus [43]:

"It is obvious that even serious investigators are today in danger to leaving the right path, if they attentively follow the newest phases in cancer theory, such as, for example, the theories of cancer immunity and of vaccination against cancer."

They concluded that "... early diagnosis and radical surgical operation is at present ... the whole art of cancer treatment. The many other methods of cure can at best only be valued as additional [today we would say: adjuvant] methods after operation or in cases of inoperable tumors".

That was the status of cancer research in 1914. The standing merit of Ehrlich, however, was his introduction of systematic experimental cancer research, and many of his experimental models are still being applied today.

Chemotherapy - Salvarsan

The last years of Ehrlich's work were dedicated entirely to chemotherapy [44], an expression that he had introduced. Many terrifying infectious diseases, especially of nonbacterial nature such as malaria, trypanosomiasis or syphilis, were obviously not amenable to serum therapy. Therefore, Ehrlich returned to chemistry. His wealth of experience with cell staining convinced him that it should be possible to systematically develop chemotherapeutic agents that would react exclusively with the chemoreceptors of the very infective agent. This was his idea of the "magic bullet".

Over a period of several years, more than 600 organoarsenic compounds were prepared, starting with atoxyl, which was already known to be effective in trypanosomiasis and syphilis, but found to be too toxic. In an unprecedented effort, hundreds of compounds were tested, most of them by Dr. Hata from Tokyo (Fig. 16), each in a large number of animals. Finally the preparation "Ehrlich-Hata 606", synthesized by Dr. Bertheim, proved to be most effective against the spirochaeta causing syphilis. Difficult and cumbersome clinical tests followed. Ehrlich himself supervised them with great care, often heavily burdened by sometimes malicious criticism, before Salvarsan (Fig. 17) was released in 1910 [45, 46]. In the same year, Fritz Haber (1868-1934) in Karlsruhe succeeded in technically converting atmospheric nitrogen into ammonia. It was a year of great advances in the mankind's fight against disease und hunger.

In Frankfurt, Paul Ehrlich received much support from two friends, the industrial chemist Arthur von Weinberg (1860-1943; Fig. 18), director of the Cassella Company, and Professor August Laubenheimer on the board of directors of the Hoechst company. With advice and action they contributed very much to the realization of Ehrlich's quite unusual plans. Ehrlich and Weinberg were close friends; both families often met at Weinberg's house, which Ehrlich once called "the birthplace of chemotherapy".

Fig. 13. Ehrlich, 1900. **Fig. 14.** Ehrlich in his laboratory in Frankfurt. **Fig. 15.** Ehrlich in his study room in Frankfurt. **Fig. 16.** Ehrlich and Dr. Hata, 1910

13
14
15
16
im Speyer-Haus
Dienstag

In appreciating Ehrlich's work on chemotherapy, it is irrelevant to note that Salvarsan could later be improved (Neo-Salvarsan). After a quarter of a century Salvarsan finally became outdated with the introduction of Penicillin. It was Ehrlich's successful strategy, however, his principle of fighting infectious diseases that was new. From that time on this principle was to be repeated in many variations, above all in the projects of the future pharmaceutical industry.

From Cell Entities to Partial Cell Functions

The idea of the fine spatial structure of the substance of all living cells with their membrane and protoplasm is found as the central theme throughout Ehrlich's work. In his Nobel lecture (1908), Ehrlich considered the many functions of the cell and said [47]:

"The time has come to analyze the biochemistry of the cell and to dismantle the conception of the cell entity into a large number of partial functions."

With this statement he advanced beyond Virchow's cellular pathology. On the basis of Pasteur's discovery of stereoisomerism, Van't Hoff had developed the stereochemistry of simple organic molecules. His ideas led Emil Fischer logically to the spatial formulae of dextrose and the other stereoisomeric sugars with five asymmetric carbon atoms. Paul Ehrlich ventured to explore the cell protoplasm with its hundreds or thousands of asymmetric centers. How delighted would he have been with today's spatial models of antibodies, his antitoxins! In 1909, he summarized his ideas and wrote [48]:

"If I have been able to advance to a certain extent toward the elucidation of the structure of protoplasm, I am only carrying out a programme that Goethe's genius already foresaw when he said:

"Had Nature, in its lifeless beginning, not been so thoroughly stereometric, how could it lastly have achieved such unpredictable and immeasurable life"? "

Fig.17. From Ehrlich's notebook: 606 formula of Salvarsan. **Fig. 18.** Arthur von Weinberg around 1910. **Fig. 19.** Ehrlich's book about work on immunology, dedicated to Dr. Friedrich Althoff and especially presented "seinem lieben E. von Behring". Inset: The exlibris on Emil von Behring. **Fig. 20.** Last photograph of Paul Ehrlich taken in 1914

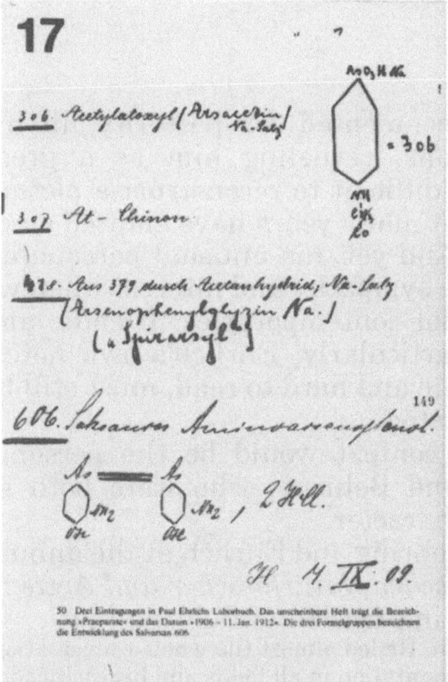

50 Drei Eintragungen in Paul Ehrlichs Laborbuch. Das unscheinbare Heft trägt die Bezeichnung »Praeparate« und das Datum »1906 – 11. Jan. 1912«. Die drei Formelgruppen bezeichnen die Entwicklung des Salvarsan 606

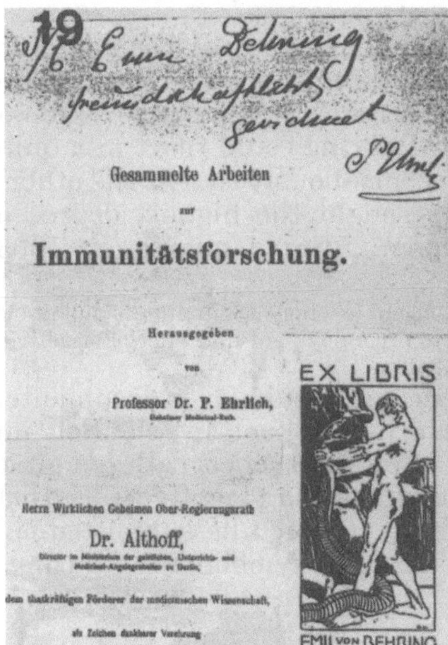

Gesammelte Arbeiten

zur

Immunitätsforschung.

Herausgegeben

von

Professor Dr. P. Ehrlich,

Geheimer Medicinal-Rath.

Herrn Wirklichen Geheimen Ober-Regierungsrath

Dr. Althoff,

Director im Ministerium der geistlichen, Unterrichts- und Medicinal-Angelegenheiten zu Berlin,

dem thatkräftigen Förderer der medicinischen Wissenschaft,

als Zeichen dankbarer Verehrung

zugeeignet.

Letzte photographische Aufnahme Ehrlichs

Science and Humanity

Paul Ehrlich's work is well documented for posterity, in his almost inexhaustible publications, revealing him as a great investigator. However, it seems difficult to reconstruct a picture of Paul Ehrlich's person now that many years have elapsed since his death on 20 August, 1915. And yet, his unusual personality deserves our full and affectionate sympathy and interest when we consider the many remarks from contemporaries, friends, and colleagues [7,10,18,38,49,50]. Particularly, Ehrlich's own notes, which are mostly in telegram style and hard to read, must still be carefully and competently evaluated.

A special chapter in this context would be the personal relationship between Ehrlich and Behring, who were both so different in their attitude and character.

Ludwig Aschoff, who met Behring and Ehrlich at the annual meeting of the *Gesellschaft Deutscher Naturforscher und Ärzte* in 1903 in Kassel, wrote about his impression of Behring [40]

"Behring is always a practical man. He led almost the whole conversation, and I must say that I admire his quick orientation in all fields and his coolheaded criticism, even against himself. Whether his theories are right does not matter to him, as long as his method leads to practical help and success ....".

Scientifically, both Ehrlich and Behring were closely related, and they cooperated for many years, especially with regard to the efficiency control of diphtheria antisera. If certain batches did not meet the necessary requirements, Behring could react in a quite unjustified manner. Each of them continuously sought the other's friendship, but they both equally sought the highest degree of personal independence and freedom. After a quarrel in 1903, Ehrlich wrote to Behring:

"I am, as you know, primarily so eager to enjoy a maximum amount of freedom that nothing can do more harm to my enjoyment and energy of activity than even the slightest constraint or obligation...".

Then, however, they became friends again, as is documented by many handwritten letters which are kept in the Behring archives in Marburg. Dr. Althoff often intervened with his great authority, understanding, and psychological sensitivity. Ehrlich dedicated an important book to him, of which he submitted one copy to Behring. See his exlibris (Fig. 19) symbolizing Behring's attitude of fighting diseases!

At Ehrlich's funeral, Behring, who survived him by only 2 years, could only say: "Dear friend, you always had a sensitive soul, and if ever we have hurt you ..., forgive us".

I would like to quote Chaim Weizmann from his autobiography *Trial and Error*, in which he describes a visit to Ehrlich in his institute in Frankfurt in 1914 [51]:

> "I have retained an ineradicable impression of Ehrlich. His figure was small and stocky, but his head was of great beauty, delicately chiseled; and out of his face looked a pair of eyes which were the most penetrating that I have ever seen - but they were eyes filled with human kindness".

This genius, who did not take care of his delicate body and who in certain periods of his life aged in a shockingly fast manner, as can be seen from many photographs (Fig. 20), this genius could not succeed without the fatherly care of the strong ones in life: the understanding teachers, Waldeyer and Frerichs, the great Althoff in the Ministry of Cultural Affairs, or his industrial friends Weinberg and Laubenheimer. On the analogy of the question "Would Mozart have had the opportunity to develop in our times", we might ask the same for Ehrlich.

In a contemplative obituary, Carl Ludwig Schleich (1859-1919), the surgeon who introduced local anesthesia, who stayed in friendly contact to Ehrlich for many years after they met in Berlin, wrote [52, 53]:

> "For thousands of years, mankind has been fighting against the threats of Nature. Is it possible for medical science to support the slow but steady progress of Nature's self protection? Can one face death and heal with Ehrlich's weapons not individually but generally, for example by studying a blood sample in the laboratory without seeing the patient face to face?"

Schleich referred to the Wassermann reaction which, when positive, now led to the automatic prescription of Salvarsan. In spite of all the satisfaction derived from achievements and further possibilities, the depressing vision appeared of an impersonal machinery which might cause the dehumanization of medicine.

It seems, that Ehrlich, who was a most affectionate and highly responsible physician in the hospital, also recognized this problem as a result of his own work. At the time of his death in August 1915 - the First World War was in full action, very much to the deep concern of scientific friends on both sides -, Ehrlich's way of thinking was certainly controversial, linking the great optimism of the nineteenth century and all the uncertainties of further development in the twentieth century.

Research is imposed on mankind; it is part of human culture. But the question arose of whether it improves, enriches, and facilitates life and quality of life as our forefathers so strongly believed, and whether science - as Paul Ehrlich saw it - will ultimately serve to diminish human suffering.

References

1. Virchow R (1849) Mitteilungen über die in Oberschlesien herrschende Typhus-Epidemie. Arch Pathol Anat Physiol Klin Med 2:143-322
2. Pasteur L (1864) Des générations spontanées. Revue des Cours scientifiques 1:257-265
3. Lister J (1867) On the antiseptic principle in the practice of surgery. Lancet 2:353-356
4. Lister J (1868) On the antiseptic treatment in surgery. Br Med J 2:53-56, 101-102, 461-463, 515, 517
5. Metschnikoff E (1933) Trois fondateurs de la médecine moderne: Pasteur, Lister, Koch, vol 1. Alcan, Paris
6. Ehrlich P (1906) Weigerts Verdienste um die histologische Wissenschaft. In: Rieder R (ed) Carl Weigert - Gesammelte Abhandlungen, vol. 1. Springer, Berlin, pp 138-141
7. Waldeyer W (1941) Paul Ehrlich. In: Paul Ehrlich - eine Darstellung seines wissenschaftlichen Wirkens. Fischer, Jena, pp 17-23
8. Ehrlich P (1877) Beiträge zur Kenntnis der Anilinfärbungen und ihrer Verwendung in der mikroskopischen Technik. Arch Mikrosk Anat 13:263-277
9. Ehrlich P (1882) Über die Färbung der Tuberkelbazillen. Dtsch Med Wochenschr 19:269
10. Michaelis L (1914) In: Paul Ehrlich - Eine Darstellung seines wissenschaftlichen Wirkens. Fischer, Jena, pp 28-36
11. Ehrlich P (1885) Das Sauerstoff-Bedürfnis des Organismus - eine farben-analytische Studie. Hirschwald, Berlin
12. Ehrlich P (1887) Über die Methylenblau-Reaktion der lebenden Nervensubstanz. Biol Centbl 6:204
13. Vant't Hoff JH (1875) La chimie dans l'espace. Rotterdam
14. Vant't Hoff JH (1874) Proposal for the extension of the formulas now in use in chemistry into space - together with a related remark on the relation between the optical rotation power and the chemical constitution of organic compounds. Utrecht
15. Walden P (1925) Vergangenheit und Gegenwart der Stereochemie. Naturwissenschaften 13:301-312
16. Freudenberg K (ed) (1933) Stereochemie. Deutike, Leipzig
17. Kussmaul AK (1885) Zum Andenken an Theodor Frerichs. Tageblatt des 58. Kongresses der Gesellschaft Deutscher Naturforscher und Ärzte, Strasbourg, 18-23 Sept 1885, pp 36-42
18. Pinkus F (1915) Paul Ehrlichs Wirken. Med Klin 1915 :1116-1117, 1143-1145

19. Behring E, Kitasato S (1890) Über das Zustandekommen der Diphterie-Immunität und der Tetanus-Immunität bei Tieren. Dtsch Med Wochenschr 16: 1113-1114
20. Ehrlich P (1891) Experimentelle Untersuchungen über Immunität. I. Über Ricin. Dtsch Med Wochenschr 17:32
21. Ehrlich P (1891) Experimentelle Untersuchungen über Immunität. II. Über Abrin. Dtsch Med Wochenschr 17:44
22. Arrhenius S (1907) Immunchemie. Akademische Verlagsgesellschaft, Leipzig
23. Dale H, Himmelweit F (eds) (1957) Immunology and cancer research, vol 2. Pergamon, London, pp 406-422 (The collected papers of Paul Ehrlich)
24. Ehrlich P (1897) Fortschritte der Medizin. In: Dale H, Himmelweit F (eds) (1957) Immunology and cancer research, vol 2. Pergamon, London, pp 93-94 (The collected papers of Paul Ehrlich)
25. Fischer E (1894) Einfluß der Konfiguration auf die Wirkung der Enzyme I. Ber Dtsch Chem Ges 27:836-847
26. Fischer E (1909) Untersuchungen über Kohlenhydrate und Fermente I (1884-1908). In: Bergmann M (ed) Springer, Berlin
27. Fischer E (1922) Untersuchungen über Kohlenhydrate und Fermente II (1908-1919). In: Bergmann M (ed) Springer, Berlin
28. Witt ON (1876) Zur Kenntnis des Baues und der Bildung färbender Kohlenstoffverbindungen. Ber Dtsch Chem Ges 9:522-527
29. Ehrlich P (1897) Über die Antitoxinwirkung. Theorie der Immunität. In: Dale H, Himmelweit F (eds) (1957) Immunology and cancer research, vol 2. Pergamon, London, pp 93-95 (The collected papers of Paul Ehrlich)
30. Heymann B (1928) Zur Geschichte der Seitenketten-Theorie von Paul Ehrlich. Klin Wochenschr 7:1257-1260, 1305-1309
31. Rieder R (ed) (1906) Carl Weigert - Gesammelte Abhandlungen, vol 1. Springer, Berlin, p 16
32. Ehrlich P (1901) Schlußbetrachtungen, vol 8. In: Nothnagel H (ed) Spezielle Pathologie und Therapie. Hölder, Vienna, p 168
33. v Behring E (1898) Allgemeine Therapie der Infektionskrankheiten. Springer, Berlin
34. Metschnikoff E. German translation: Meyer J (1902) Immunität und Infektionskrankheiten. Fischer, Jena, p 444
35. Ehrlich P, Morgenroth J (1899) Zur Theorie der Lysinwirkung. Berl Klin Wochenschr 1899:150-164, 196-204, 213-223, 234-245, 256-277
36. Aschoff L (1902) Ehrlichs Seitenkettentheorie und ihre Anwendung auf die künstliche Immunisierungsprozesse. Z Allg Physiol 1B:69-248
37. Ehrlich P (1899) Mode of Action and Mechanism of Production of Antitoxins. In: Dale H Himmelweit F (eds) (1957) Immunology and cancer research, vol 2. Pergamon, London, pp 173-177 (The collected papers of Paul Ehrlich)
38. Lazarus A (1922) Meister der Heilkunde. Rikola, Vienna, pp 75, 81
39. Jensen CO (1907) Z Krebsforsch 5: 164
40. Aschoff L (1966) Briefe. Schulz, Freiburg, p 153
41. Loeb L (1950) Bibliography of writings of Dr. Leo Loeb from 1896 to 1949. Arch Pathol (Chicago) 6:661-675
42. Ehrlich P (1909) Über den jetzigen Stand der Carzinomforschung. In: Dale H, Himmelweit F (eds) (1957) Immunology and cancer research, vol 2. Pergamon, London, pp 550-562 (The collected papers of Paul Ehrlich)
43. Wolff J (1914) Die Lehre von der Krebskrankheit. Fischer, Jena, p 576

44. Dale H, Himmelweit F (eds) (1960) Chemotherapy, vol 3. Pergamon, London (The collected papers of Paul Ehrlich)
45. Neisser A (1910) Moderne Syphilistherapie. 82. Kongreß der Gesellschaft Deutscher Naturforscher und Ärzte, Königsberg, 18.-24. Sept. 1910 (Verhandlungsprotokolle) 1:172-183
46. Ehrlich P (1911) Über Salvarsan. Kongreß der Gesellschaft Deutscher Naturforscher und Ärzte, Karlsruhe, 24.-29. Sept. 1911 (Verhandlungsprotokolle) 1:299-315
47. Ehrlich P (1908) Über Partialfunktionen der Zelle. In: Dale H, Himmelweit F (eds) (1960) Chemotherapy, vol 3. Pergamon, London, pp 171-182 (The collected papers of Paul Ehrlich) 48. Ehrlich P (1909) Beiträge zur experimentellen Pathologie und Chemotherapie. In: v Weinberg A (1914) Paul Ehrlich - Eine Darstellung seines wissenschaftlichen Wirkens. Fischer, Jena, pp 3-13
49. Marquardt M (1951) Paul Ehrlich. Springer, Berlin
50. Bäumler E (1979) Paul Ehrlich - Forscher für das Leben. Societät, Frankfurt
51. Weizmann C (1949) Trial and error. Hamilton, London, p 180
52. Schleich CL (1915) Paul Ehrlich - ein Nekrolog. Die neue Rundschau 2:1409-1419
53. Schleich CL (1931) Besonnte Vergangenheit - Lebenserinnerungen 1859-1919. Rowohlt, Berl, pp 305-323

Soluble Lymphocyte Receptors

M. Schlesinger

The Paul Ehrlich Center for the Study of Normal and Leukemic White Blood Cells, The Hebrew University-Hadassah Medical School, Jerusalem, Israel

Almost a century ago, Paul Ehrlich elaborated his "side chain" theory to explain the mechanism of antibody formation [1]. According to his theory, cells capable of producing antibodies display on their surface side chains fitted to react with antigens like male and female screws or like a key with the corresponding lock. Ehrlich further proposed that the binding of antigen to the side chains provoked vigorous synthesis of further side chains. In his view, anibodies represent side chains produced in excess and, therefore, pushed off the cytoplasm, coming to exist in a free state. Ehrlich's concept that immunocompetent cells possess specific antigen receptors forms the basis of modern theories of clonal selection in immunology.

In full agreement with Ehrlich's vision, immunoglobulin molecules were found to exist both as cell surface antigen receptors in the cell membrane and in solution as secreted antibodies. The molecular biology underlying the generation of antibody was elucidated and the biochemical processes underlying the formation of various immunoglobulin molecules have been unravelled. The μ immunoglobulin chains present in the cell membrane and the μ chains in the secreted IgM molecules were found to differ. The membrane-associated chains have distinct hydrophobic C terminal segments and are longer than the secreted μ chains [2,3]. Rogers et al. [4] reviewed four different models which could account for the differences between membrane and secreted μ chains:

- The secreted μ chain could be produced by proteolytic cleavage of the membrane associated μ chains so as to remove the hydrophobic C terminus.
- The constant μ gene may have its 3' coding region rearranged during B-cell differentiation.
- The membrane and secreted μ chains may be encoded by two different constant μ genes.

- The membrane and secreted μ chains may be generated by alternative RNA splicing.

Rogers et al. [4,5] ruled out the first three mechanisms and provided evidence that the expression of membrane μ chains and secreted μ chains result from developmentally regulated RNA processing. In addition to the expression of immunoglobulin molecules in B-cells, lymphocytes possess a whole array of receptors on their cell surface, some of which are characteristic for distinct lymphocyte subsets. The role of various constituents of the cell surface of lymphocytes in controlling their functions, in their interactions with other cells, and in signal transduction are under intensive investigation. Some of the constituents of the cell surface membrane of lymphocytes can be shown to become released from the cells into the extracellular milieu both in vitro and in vivo. The aim of the present talk is to discuss some of the problems related to the release of lymphocyte surface receptors, other than immunoglobulins. No exhaustive review of soluble lymphocyte receptors is intended. Illustrative examples will be given for the release of lymphocyte receptors, an extension of Ehrlich's concept that functional side chains are pushed off the cell surface as part of the activity of the cell. The mechanisms of release and the possible role of soluble receptors in the immune system will be discussed.

Soluble Class I MHC Molecules

Antigenic determinants of the major histocompatibility complex (MHC) play a vital role in determining the immune specificity and associative recognition by T-lymphocytes. Class I MHC molecules affect predominantly the immune reactivity of the CD8+ T-lymphocyte subset. Is there any evidence for the release of soluble class I molecules in a way that could affect the immune system [6]?

The class I MHC molecules are composed of a heavy and a light chain, encoded by genes located on different chromosomes. The heavy chain is a glycosylated peptide, spanning the cell membrane, with a molecular weight of 40 000 - 50 000. The light chain, β_2-microglobulin (β_2M), has a molecular weight of 12 000 and is noncovalently associated with the heavy chain on the extracellular side of the plasma membrane.

A number of papers have demonstrated soluble class I MHC peptides. Kress et al. [7] found that mouse liver cells synthesize and secrete a soluble class I related protein (Q10). Within the region of the gene corresponding to the transmembrane domain, a number of substitutions of hydrophobic residues rendered the encoded protein incapable of inserting into the plasma membrane. The Q10 gene also lacked the intracellular domain. This unusual class I molecule could be demonstrated not only in liver cells, but also in the serum of certain strains of mice. Variations in the expression of the Q10 protein in the serum among different strains of mice, paralleled strain variations in the ribonucleic acid (RNA) transcripts of this protein in the liver.

The finding that the liver has a unique capacity to secrete soluble MHC peptides seemed to correlate with another unique feature of the liver. Experimental liver transplants were found not to be rejected, even in the case of histoincompatibility between the donor and the recipient. Moreover, liver allografts into previously sensitized recipients not only resisted rejection, but even induced in the hosts a state of nonreactivity, characteristic of immune tolerance [8]. Since shortly after liver engraftment class I molecules of donor origin could be found in the serum of the recipients, it has been suggested that the secretion of soluble, truncated class I antigens by liver transplants may contribute to their survival in incompatible hosts. According to this contention, soluble class I proteins may interfere with the recognition process of cytotoxic T-lymphocytes and block their reactivity. While the suggestion that soluble class I molecules may be tolerogenic is attractive, it has yet to receive direct experimental support. Indeed, Mann et al. [9] demonstrated that mice who had soluble Q10 in their serum were fully capable of mounting a cytotoxic reaction against Q10 antigenic determinants anchored, by genetic engineering, to the cell surface of L-cells. Thus, the role of soluble Q10 in the induction of tolerance is at present in doubt.

Important information on the release of Qa region class I antigens was obtained by Robinson [10] in studies of mouse lymphocytes subjected to mitotic stimulation in vitro. Soluble molecules carrying the Qa-2 and Qb-1 specificities were found in supernates of T-lymphocytes stimulated by concanavalin A and of lipopolysaccharide-stimulated B-cells. His results indicated that these class I molecules are released by two different mechanisms. Qb-1 seemed to be secreted directly, as a 42000 molecular weight peptide, and could not be detected in the cell membrane. In contrast, Qa-2 became first expressed on the cell membrane, and only subsequently was released into the supernatant fluid, as a 37000 molecular weight peptide. Thus, while the Qb-1 molecule is secreted by the cells, the release of Qa-2 from the cell membrane seems to involve cleavage of the anchor of the molecule to the membrane.

Not only Qa region class I molecules are released from the surface of murine lymphocytes. Emerson et al. [11] demonstrated that both H-2D and H-2K gene products of the murine MHC system are shed in vitro from the membrane of unstimulated spleen cells. MHC determinants were found to be shed from the surface of murine spleen cells as particles ranging in molecular weight from 3×10^5 to 2×10^6. Evidence was presented that these antigenic determinants were shed in association with local membrane lipids, as small membrane derived vesicles. In contrast, cell surface IgM molecules were shed without detectable associated lipids. In some H-2 haplotypes, the shedding and synthesis of the glycoproteins encoded by the H-2D and H-2K loci proceeded at different rates. Thus in the CBA/J spleen cells, H-$2K^k$ antigens were found to be rapidly turned over and shed, whereas the turnover and shedding of the H-$2D^k$ antigens were extremely slow. A striking correlation was found between the variations in shedding rates of H-2K and H-2D antigens in different H-2 haplotyes and variations in H-2 restricted T-cell responses to modified or virus infected target cells. Rapidly shed H-2 molecules, whether coded by the K or D regions, appeared to constitute preferred vehicles for T-cell stimulation. Thus, the shedding of H-2 molecules may be instrumental in the presentation of foreign membrane-bound antigens to the immune system.

Soluble class I MHC molecules of the specificity human leukocyte antigen (HLA)-Aw24 (A9) were detected in human serum [12]. The work of Krangel [13] indicated that the HLA-Aw24 gene is more prone than any other class I HLA gene to produce messenger (m)RNA, in which the transmembrane exon is removed by alternate splicing. Krangel studied the in vitro release of 10 different HLA-A specificities and of 13 HLA-B specifities by a series of human cell lines and by mitogen-activated lymphocytes. The only HLA specificity which was released at a high rate was that of HLA-Aw24.

The light chain of class I MHC antigens, the β_2M, has been detected in all biological fluids and in urine [14, 15]. Most of the β_2M in the serum is free, and only very small amounts of β_2M (2%-5%) seem to be associated with the heavy chain of class I MHC. The level of β_2M in the serum is controlled by two mechanisms. The first of these is the glomerular filtration and tubular reabsorption of β_2M in the kidney. Blood levels of β_2M rise as glomerular filtration rate falls, while impairment of the reabsorption process in the kidney tubules leads to loss of β_2M in the urine and to decreased serum levels. The second mechanism that determines the level of β_2M in the serum is the rate of synthesis and release of β_2M from cells. Elevated levels of serum β_2M are found in inflammatory diseases involving polyclonal lymphocyte activation [14, 16]. These include rheumatoid arthritis, Sjögren's disease, systemic lupus erythematosus, sarcoidosis, Crohn's disease, and angioimmunoblastic lymphadenopathy. β_2M serum levels may also be increased in lymphoproliferative diseases and in some forms of cancer, although somewhat conflicting results were obtained in different studies [17, 18]. Low levels of β_2M where shown to be shed by lymphocytes kept in culture. Mitogenic stimulation of the lymphocytes markedly increased the rate of release of β_2M into the medium [19, 20], most of which is not associated with heavy class I molecules [10].

Soluble Interleukin-2 Receptor

Activated T-lymphocytes not only produce interleukin-2 (IL-2), but also express IL-2 receptors (IL-2R) on their cell surface [21]. IL-2R are composed of an α chain which binds IL-2 with low affinity and a β chain which binds IL-2 with high affinity. IL-2 binds with highest affinity to a receptor formed by the association of the α and β chains. Activated T-lymphocytes were shown to release the α chain of the IL-2R in a soluble form (sIL-2R) [22]. Activation of human peripheral blood mononuclear cells with various stimuli induced the expression of cell associated IL-2R and in parallel elicited the formation of sIL-2R [22]. sIL-2R is released by both activated T- and B-cells. Some stimuli elicited rapid changes while other stimuli resulted in a more gradual increase in both cellular and soluble IL-2R. All cells expressing the α chains on their surface, either transiently or constitutively, seem to be capable of releasing it, probably through proteolytic cleavage at the cell surface. The cell surface α chain of the IL-2R has a molecular weight of about 45 000, as a result of loss of the transmembrane and intracytoplasmic domains by proteolytic cleavage [23]. Increased levels of sIL-2 in the serum were noted in a large number of disorders associated with cellular activation including autoimmune diseases [24], lymphomas [22, 25], hairy cell leukemia [26], and transplantation rejection [27].

Soluble CD8

On the cell surface CD8 determinants constitute dimers, with a molecular weight of about 60 000-70 000, while the soluble nonreduced form of CD8 found in the serum had a molecular weight of 52 000-54 000 [28, 29]. The level of CD8 in normal control sera was about 290 U/ml [30]. High concentrations of soluble CD8 molecules were found in the sera of patients with CD8+ acute lymphoblastic T-cell leukemia [28]. In the sera of individuals with infectious mononucleosis there is a tremendous elevation of the level of soluble CD8 (mean of about 7 900 U/ml [30]. This increase paralleled the elevation in the number of activated CD8+ lymphocytes in the blood of these patients. Increased levels of soluble CD8 were also found in the sera of patients with lymphoproliferative diseases affecting CD8-negative cells. Elevated serum CD8 levels were found to be a poor prognostic sign in children with acute lymphoblastic leukemia and in non-Hodgkin's lymphoma [31]. Among children with Hodgkin's disease, the level of serum CD8 was significantly higher among patients with advanced disease than in those who had less extensive disease. Pui et al. [32] suggested that the increased CD8 serum level reflects enhanced suppressor T-cell activity, which may compromise the host's anti-tumor immunity, which in turn would lead to more aggressive disease.

Soluble CD2

Soluble CD2 molecules (previously called E-receptors) were demonstrated in the serum by various techniques. Mendes et al. [33] detected soluble CD2 by an electroimmunodiffusion technique, using a xenogeneic antiserum raised in sheep by immunization with E-receptor-coated autologous red blood cells. Elevated levels of CD2 were detected in the serum of patients with cancer, lymphoproliferative diseases, uremia [34], and aplastic anemia [35]. Oh et al. [36] developed a radioimmunoassay for the detection of soluble CD2. Using this method they demonstrated the release of CD2 molecules by activated T-lymphocytes [36]. High levels of CD2 were found in ascitic effusions in cancer and in the serum of leukemic patients.

We have recently analyzed the concentration of CD2 in the serum by an enzyme-linked immunosorbent assay (ELISA), using a monoclonal CD2 antibody [37]. In ELISA tests for repetitive determinants, considerable levels of CD2 were detected in the sera of only 14% of 63 normal controls tested. In contrast, employing an ELISA test for the detection of single CD2 determinants, the sera of all individuals tested were found to contain soluble CD2. Sera of individuals with activated T-cells had significantly elevated concentrations of serum CD2. Some patients with acute lymphoblastic leukemia (ALL) also had increased CD2 levels. In one patient with ALL who had low levels of soluble CD2 fluctuations of soluble CD2 were found to correlate with changes in the proportion of T-lymphocytes. Thus, soluble serum CD2 may reflect both the number of T-cells and their state of activation.

Regulation of the Level of Soluble Receptors in the Serum

The level of various soluble receptors in the serum seems to reflect a complex interplay of numerous factors. One of the factors that would reduce the level of soluble receptors found in the serum is their rate of loss in the kidney. Soluble receptors detected in the urine include β_2M [14], HLA [12], CD14 [38] and tumor necrosis factor (TNF) [39]. Another mechanism that could control the serum level of soluble receptors is their specific interaction with ligands. Thus, it can be envisaged that, to some degree, the serum level of soluble CD8 molecules may be controlled by their interaction with class I antigenic determinants present on practically all cells. Similarly, the serum level of soluble CD2 molecules in the serum could be reduced by their interaction with the widespread lymphocyte function associated antigen (LFA)-3 determinants [40]. Yet another mechanism for the removal of soluble receptors from the circulation may be their interaction with autoantibodies. We have suggested that the frequent occurrence of autoantibodies to determinants of the Thy-1 antigen [41] may play a role in the elimination of soluble Thy-1 [42]. Moreover, since soluble Thy-1 may be immunosuppressive [43], autoantibodies to Thy-1 may prevent the interference of soluble Thy-1 with the immune response [42]. Autoantibodies to β_2M were detected in the serum of patients with systemic lupus erythematosus [42]. Revillard et al. [44] suggested that such autoantibodies may be part of the normal immune response to foreign antigens associated with histocompatibility antigens. It will be of interest to see whether autoantibodies may play a role in controlling the serum level of other soluble receptors.

Increased serum levels of soluble receptors may result either from increased release from the surface of intact lymphocytes, or from accelerated destruction of lymphocytes. Like other living cells, components of the cell membrane of cells of the lymphoreticular system constantly shed surface constituents into their surroundings. Shedding constitutes a physiological, continuous release of macromolecules derived from the surface membrane of living cells, as part of the turnover of the surface membrane [45]. Greater quantities of particular component may be shed when the number of cells possessing this component is increased. The proliferation of lymphocytes may increase the overall amount of distinctive molecules shed from their cell surface and in this way elevate the amount being detected in the serum. Thus, the proliferation of CD8+ lymphocytes in infectious mononucleosis [30] or in T-cell leukemia [28] may lead to increased serum levels of soluble CD8. Following lymphocyte stimulation various cell surface constituents are released at an accelerated rate. Some receptors such as IL-2R become expressed on the cell surface only following stimulation [21] and are shed shortly thereafter [22]. Other receptors shed from the membrane of stimulated lymphocytes, like CD8 [30], β_2M [19, 20] and CD2 [36] are expressed on the cells prior to their stimulation. It seems that the shedding of these surface receptors parallels the appearance of markers of activation of lymphocytes [30]. The accelerated release of cell surface receptors from activated cells may reflect an increased concentration on the cell surface. It is also possible that activation of lymphocytes leads, to a limited extent, to the production of receptors with incompletely developed transmembrane anchors, as a result of alternate splicing of mRNA [10]. Cell activation may also increase the activity of enzymes that could split off surface receptors. CD14 is a molecule whose anchor to the cell membrane can be digested with phosphatidylinositol-specific phospholipase C (PI-PLC) [46]. The loss of CD14 from the surface of activated monocytes and macrophages [47] could be partially attributed to digestion by PI-PLC. Kubler et al. [48] have shown that protein kinase present on the surface on intact cells can be released from the cell surface into the supernate by interaction with the substrate. Concomitantly, the cells were depleted of their surface protein kinase. The release of surface protein kinase was a rapid process, demonstrable within minutes. The restoration of surface protein kinase was a more gradual process, taking a few hours.

Soluble receptors may be released into the circulation not only from living cells but also from dead cells. Volk et al. [49] observed that the administration of immunosuppressive doses of monoclonal antibodies directed against IL-2R resulted in a twelvefold increase of sIL-2R in the serum, while immunosuppressive doses of monoclonal L3T4 antibodies had no such effect. They concluded that the increase of sIL-2R probably reflects the destruction of IL-2R bearing cells by the anti-IL-2R target immunotherapy.

Soluble Receptors in Acquired Immunodeficiency Syndrome

The serum of human immunodeficiency virus (HIV)-infected individuals contains increased concentrations of a number of soluble receptors. It is not quite clear, however, what mechanisms underly this change. Numerous studies have shown a significant increase of sIL-2R in the serum of HIV-infected subjects. Some studies indicated that the level of this receptor increases with the progression of the disease [50, 51, 52]. In some studies [50, 51] the increase sIL-2R level was inversely correlated with the numer of CD4+ lymphocytes. Prince et al. [53] suggested that this may reflect a vicious circle caused by activation of CD4+ cells by intercurrent infections. Activation of CD4+ cells may amplify HIV-1 replication with subsequent acceleration of CD4 cell destruction, and reduction of their number [53]. Activation may in parallel result in an increased serum sIL-2R level. Although Honda et al. [50] also found inverse correlation between the serum level of sIL-2R and the number of CD4 cells, they expressed doubts whether activation of T-cells was the source for elevated serum levels of sIL-2R in HIV-infected and in acquired immunodeficiency syndrome (AIDS) patients. Peripheral blood mononuclear cells from AIDS patients failed to secrete increased amounts of sIL-2R. Moreover, while T-cell lines infected with human T-cell lymphoma virus (HTLV)-1 and -II, were found to produce large quantities of sIL-2R; HIV-infected T-cell lines produced relatively small amounts of sIL-2R. Honda et al. [50] raised the possibility that the increase of sIL-2R may result from destruction of IL-2R bearing cells similar to that documented by Volk et al. [49]. Another possibility they considered was that HIV infection may impair the clearance of sIL-2R from the serum by disturbing the function of the liver or kidney. Lang et al. [51] confirmed that the serum level of sIL-2 was elevated in HIV infection, but found no correlation with either the number of CD4+ peripheral blood leukocytes or that of activated T-cells. They, too, concluded that the sIL-2R in seropositive subjects did not emanate from activated T-cells and raised the possibility that activated B-cells or monocytes could lead to elevated sIL-2R serum levels.

Another receptor the serum level of which is raised in HIV-infected individuals and in AIDS patients is $ß_2$M [51, 52, 54]. Franzetti et al. [54] noted that AIDS patients had higher concentrations of $ß_2$M than subjects in earlier stages of infection. Again, it is not quite clear whether the increased level of this protein reflects lymphocyte activation, cell destruction, or perhaps impairment of renal function.

In preliminary studies on the level of CD2 in the serum of HIV-infected individuals, two phases were noted [37]. While low levels of CD2 were detected in 14% of the normal controls tested, in none of the asymptomatic HIV infected subjects tested could soluble CD2 be found. Elevated concentrations of CD2 were found in the sera of 3 out of 12 AIDS patients. In one of the AIDS patients, episodes of T-cell destruction and reduction in the number of CD4+ cells were clearly correlated with increased levels of soluble CD2 levels in the serum. Thus, the decline in the serum level of soluble CD2 in early phases of HIV infection may reflect impaired T-cell stimulation, while its increased level in some of the AIDS patients may indicate cell destruction.

Do Soluble Receptors Have a Function?

There are three a priori possibilities for the effects that soluble receptors could have within the immune system [55].

- Soluble receptors could exert an inhibitory effect by competing with cell-bound receptors for agonists. In most systems, the possible inhibitory role of soluble receptors is usually discussed.
- Another, diametrically different possibility is that soluble receptors may augment and supplement the reactivity exerted by cell-associated receptors. Soluble receptors may bind specific ligands at a distance from the receptor-bearing cells. Solubility of the receptors may thus enable specific interactions beyond the reach of cells expressing receptors on their surface.
- The third possibility is that, unlike cell-associated receptors, soluble receptors may not have a noticeable function. Thus, the low affinity of soluble receptors to ligands and the lack of coupling to transmembrane signalling components may render them inactive.

It should be emphasized that the concentration of soluble receptors in the vicinity of releasing cells would be considerably higher than in the serum. Therefore, whatever their function, receptors released from the cell surface would be most effective in the immediate microenvironment of their site of release. This is particularly pertinent in those instances in which the cell surface receptors are released in association with relatively large membrane fragments [11, 56]. Indirect evidence supporting the augmentation of the immune response by receptors released from the cell membrane was presented by Emerson et al. [11]. The capacity of either H-2D or H-2K determinants of various H-2 haplotypes to control immune responsiveness was found to be correlated with their rate of shedding from the cell membrane. Similarly, I-A molecules, which play an important role in the immune response, were shown to turn over and shed at a faster rate than I-E molecules [56].

Doyle et al. [57] have recently studied the effect of HIV-1 components and of soluble CD4 (sCD4) on the interaction of radioactively labeled Raji B-lymphocytes expressing class II MHC determinants with monolayers of CD4+ cells. The gp 120 component of the HIV-1 virus completely abrogated this interaction, probably by competitive binding to the CD4 molecule on the CD4+ monolayer. The addition of sCD4 alone enhanced, rather than blocked, the binding of B-lymphocytes to CD4 bearing cells. The authors suggest a number of alternative mechanisms for this enhancement.

- The interaction of sCD4 with cell surface class II molecules on lymphocytes could induce clustering of class II molecules and thereby increase their affinity.
- Interaction of sCD4 with membrane-bound CD4 enhances polyvalent interactions.
- sCD4 removes a natural inhibitor of the interaction between the CD4 and class II MHC molecules.

Numerous papers indicate that soluble lymphocyte receptors can interfere with the immune response. sIL-2 R was shown to be capable of binding IL-2 efficiently [58]. Therefore, high levels of sIL-2R can prevent IL-2 from reacting with target cells and interfere with the triggering of T-cells by IL-2. In the serum of patients with hairy cell leukemia, highly elevated levels of sIL-2 were found [26], probably produced by the tumor cells [59]. It was suggested that sIL-2R may, in part, be responsible for the immune impairment observed in this disease [26]. In rheumatoid arthritis, T-cells show a distinctive deficiency in their response to IL-2, and synovial exudate fluid was found to inhibit the stimulation of T-cells by IL-2. Symons et al. [60] demonstrated that serum levels of sIL-2 R were closely correlated with disease activity. The ability of the synovial fluid from patients with rheumatoid arthritis to inhibit the response to IL-2 was correlated with their sIL-2 content. Moreover, fractionation of the synovial fluid by gelfiltration demonstrated that the inhibitor of IL-2 response and the sIL-2R co-eluted in the same fraction, consistent with the notion that sIL-2R plays a central role in the downregulation of the response to IL-2.

CD2 molecules on the membrane of human T-cells regulate their immune reactivity [40]. Antibodies against a single epitope of the CD2 molecule inhibited various T-cell functions, whereas the concomitant exposure of T-cells to antibodies directed against two distinct epitopes of CD2 elicited mitotic triggering [40, 61]. The natural ligand, to which the CD2 molecule binds, seems to be the LFA-3 molecule [40], and soluble CD2 was shown to interfere with this interaction [62]. Immunosuppression was attributed to soluble CD2 present in the serum of various groups of patients [33-35]. The immunosuppressive potential of soluble CD2 was demonstrated by Oh and Moolten [63]. Studying an immunosuppressive factor found in the ascitic fluids of ovarian cancer they determined that it constituted soluble CD2.

Concluding Remarks

Ehrlich's side chain theory laid the ground for the concept of lymphocyte stimulation and antibody secretion. It also constituted the basis for the concept of clonal selection in immunology. In addition to immunoglobulin molecules, lymphocytes release a whole array of other side chains, i.e., other receptors involved in the various immune functions of the lymphocyte. There remains a lot to be learned about the mechanisms involved in the release of these molecules, whether by shedding, secretion, or yet other processes. Many aspects of the regulation of the release of soluble receptors and their handling remain to be elucidated. The appearance of soluble receptors in the serum is of clinical interest, but the full significance of changes in their level remains unclear. With the recent advances in molecular biology, the administration of soluble receptors has come of age. Already, preparations of soluble CD4 are administered to HIV-infected individuals in an attempt to block the interaction of the HIV virus and the CD4 molecule on T-lymphocytes. In the near future it is to be expected that soluble receptors for TNF and for IL-2 will become available for pioneering attempts to block autoimmune disturbances. Conversely, in those conditions where soluble receptors will be shown to have an adverse effect on the immune system, it is to be expected that procedures will be developed to remove them from the circulation. In short, Ehrlich's concept of the importance of side chains in immunology can be safely assumed to yield further exciting benefits to mankind in the days to come.

Acknowledgments. This project was supported in part by the Goldhirsch Foundation. The collaboration of Dr. R. Rabinowitz and Mrs. R. Hadar in the studies carried out in our laboratory is gratefully acknowledged.

References

1. Ehrlich P (1957) Gesammelte Arbeiten in vier Bänden, vol 2, Immunitätslehre und Krebsforschung. Pergamon, London
2. Melcher U, Uhr JW (1973) An electrophoretic difference between surface and secreted IgM of murine splenocytes. Exp Med 138:1282-1287
3. Bergmann Y, Haimovich J (1978) B lymphocytes contain three species of μ chains. Eur J Immunol 8:876-880
4. Rogers J, Early P, Carter C, Calame K, Bond M, Hood L, Wall R (1980) Two mRNAs with different 3' ends encode membranebound and secreted forms of immunoglobulin μ chain. Cell 20:303-312
5. Early P, Rogers J, Davis M, Calame K, Bond M, Wall R, Hood L (1980) Two mRNAs can be produced from a single immunoglobulin μ gene by alternative RNA processing pathways. Cell 20:313-319
6. Gussow D, Pleogh H (1987) Soluble class I antigens: a conundrum with no solution. Immunol Today 8:220-222
7. Kress M, Cosman D, Khoury G, Jay G (1983) Secretion of a transplantation related antigen. Cell 34:189-196
8. Kamanada N, Davies HffS, Roser BJ (1981) Fully allogeneic grafting and the induction of donor-specific unreactivity. Transplant Proc 13:837-841
9. Mann DW, Stroynowski I, Hood L, Forman J (1987) Cytotoxic T lymphocytes from mice with soluble class I Q10 molecules in their serum are not tolerant to membrane bound Q10. J Immunol 138:240-245
10. Robinson PJ (1987) Two different biosynthetic pathways for the secretion of Qa region-associated class I antigens by mouse lymphocytes. Proc Natl Acad Sci USA 84:527-531

11. Emerson SG, Murphy DB, Cone RE (1980) Selective turnover and shedding of H-2K and H-2D antigens is controlled by the major histocompatibility complex. Implications for H-2 restricted recognition. J Exp Med 152:783-795
12. Vincent C, Revillard JP, Betuel H (1976) Purification of HLA antigens from urine. Transplantation 22:500-507
13. Krangel MS (1986) Secretion of HLA-A and B antigens via an alternative RNA splicing pathway. J Exp Med 163:1173-1190
14. Karlsson FA, Wibell L, Evrin PE (1980) $ß_2$-Microglobulin in clinical medicine. Scand J Lab Invest 40 [Suppl 154]:27-37
15. Bernier GM (1980) $ß_2$-Microglobulin: structure, function and significance. Vox Sang 38:323-327
16. Forman DT (1982) Beta-2 microglobulin - an immunogenetic marker of inflammatory and malignant origin. Ann Clin Lab Sci 12:447-452
17. Di Giovanni S Valenti G Carducci P, Giallonardo P (1989) Beta-2 microglobulin is a reliable tumor marker in chronic lymphocytic leukemia. Acta Haematol 81:181-185
18. del-Giglio A, Franco EL, Torloni H, Marques LA, Bretani MM, Arap W, Macchione M, Chammas R (1989) Tumor and serum beta-$_2$-microglobulin expression in women with breast cancer. Am J Clin Pathol 92:339-342
19. Bernier GM, Fanger MW (1973) Synthesis of $ß_2$-microglobulin by stimulated lymphocytes. J Immunol 109:407-409
20. Welsh KL, Dorval G, Nilsson K et al. (1977) Quantitation of $ß_2$M and HLA on the surface of human cell. II. In vitro cell lines and their hybrids. Scand J Immunol 6:265-271
21. Waldmann T (1986) The structure, function and expression of interleukin-2 receptors on normal and malignant lymphocytes. Science 232:727-732
22. Rubin LA, Kurman CC, Fritz ME, Biddison WE, Boutin B, Yarchoan R, Nelson DL (1985) Soluble interleukin 2 receptors are released from activated human lymphoid cells in vitro. J Immunol 135-3172-3177
23. Robb RJ, Kutney RM (1987) Structure-function relationships for the IL-2 receptor system IV. Analysis of the sequence and ligand-binding properties of soluble Tac protein. J Immunol 139:855-682
24. Manoussakis MN, Papadopoulos GK, Drosos AA, Moutsopoulos HM (1989) Soluble interleukin 2 receptor molecules in the serum of patients with autoimmune diseases. Clin Immunol Immunopathol 50:321-332
25. Pui C-H, IP SH, Kung P, Dodge RK, Berard CW, Crist WM, Murphy SB (1987) High serum interleukin-2 receptor levels are related to advanced disease and a poor outcome in childhood non-Hodgkin's lymphoma. Blood 70:624-628
26. Steis RG, Marcon L, Clark J, Urba W, Longo DL, Nelson DL, Maluish AE (1988) Serum soluble IL-2 receptor as a tumor marker in patients with hairy cell leukemia. Blood 71:1304-1309
27. Colvin RB, Fuller TC, MacKeen L, Kung PC, Ip SH, Cosimi AB (1987) Plasma interleukin 2 receptor levels in renal allograft recipients. Clin Immunol Immunopathol 43:273-276
28. Fujimoto J, Levy S, Levy R (1983) Spontaneous release of the Leu2 (T8) molecule from human T cells. J Exp Med 159:752-766
29. Fujimoto J, Stewart SJ, Levy R (1984) Immunochemical analysis of the released Leu2 (T8) molecule. J Exp Med 160:116-124
30. Tomkinson BE, Brown MC, Ip SH, Carrabis S, Sullivan JL (1989) Soluble CD8 during T cellactivation. J Immunol 142:2230-2236

31. Pui C-H, Ip SH, Dodge RK, Carrabis S, Brown M, Crist WM et al. (1988) Serum levels of CD8 antigen in childhood lymphoid malignancies: A possible indicator of increased suppressor cell activity in poor-risk patients. Blood 72:1015-1021
32. Pui C-H, Ip SH, Thompson E, Dodge RK et al. (1989) Increased serum CD8 antigen level in childhood Hodgkin's disease relates to advanced stage and poor treatment outcome. Blood 73:209-213
33. Mendes NF, Bernd LAG, Cintra HHA, Mendes CMF, Praca CL, Moura NC (1982) Human T lymphocyte receptor for sheep erythrocytes: characterization of a specific antiserum and its application in the detection and quantitation of the receptor in soluble form. Cell Immunol 72:143-150
34. Moura NC, Longo IM, Bernd LAG, Mendes NF (1983) Quantitation of the soluble receptor of human T-lymphocytes for sheep and erythrocytes by electroimmunodiffusion in the serum of patients with cancer, uremia and leprosy. Experientia 39:306-308
35. Falcao RP, Longo IM, Moura NC, Mendes NF (1984) Quantification of the soluble receptor of human T-lymphocytes for sheep erythrocytes in the serum of patients with aplastic anaemia. J Clin Lab Immunol 13:141-143.
36. Oh SK, Lapenson D, Morgan AC (1985) Quantification of soluble E-receptor in the serum of patients with various diseases and its accompanying in vitro immunosuppression in neoplasia. Scand J Immunol 22:51-60
37. Schlesinger M, Afik O, Loves O, Maayan S, Rabinowitz R (1990) Changes in the level of soluble CD2 molecules in the serum of HIV-infected individuals. Cancer Detect Prevent 14:347-351
38. Bazil V, Horejsi V, Baudys M, Kristofova H, Strominger JL, Kostka V, Hilgert I (1986) Biochemical characterization of a soluble form of the 53 kDa monocyte surface antigen. Eur J Immunol 16:1583-1589
39. Engelmann H, Novick D, Wallach D (1990) Two tumor necrosis factor-binding proteins purified from human urine. J Biol Chem 265:1531-1536
40. Springer TA, Dustin ML, Kishimoto TK, Marlin SD (1987) The lymphocyte function-associated LFA-1, CD2 and LFA-3 molecules: cell adhesion receptors of the immune system. Ann Rev Immunol 5:223-252
41. Thomas DB, Giguere U, Graham CM, Oliveira OLP (1986) Autoimmunity to Thy-1. Eur J Immunol 16:40-47
42. Schlesinger M (1989) The function of Thy-1. In: Reif AE, Schlesinger M (eds) In: Cell surface antigen THY-1; immunology, neurology, and therapeutic applications. Dekker, New York, pp 589-606
43. Freimuth WW, Miller HC, Esselman WJ (1979) Soluble factors containing Thy-1 antigen shed from lymphoblastoid cells modulate in vitro plaque-forming cell response. J Immunol 123:201-203
44. Revillard J-P, Vincent C, Rivera S (1979) Anti-β_2-microglobulin lymphocytotoxic autoantibodies in systemic lupus erythematosus. J Immunol 122:614-618
45. Doljanski F (1982) Cell surface shedding. In: Horowitz M (ed) The glycoconjugates, vol IV. Academic, New York, pp 155-187
46. Bazil V, Baudys M, Hilgert I, Stefanova I, Low MG, Zbrozek J, Horejsi V (1989) Structural relationship between the soluble and membrane-bound forms of the human monocyte surface glycoprotein CD14. Mol Immunol 26:657-662

47. Ugolini V, Nunez G, Smith RG, Stastny P, Capra JD (1980) Initial characterization of monoclonal antibodies against human monocytes. Proc Natl Acad Sci 77:6764-6768
48. Kübler D, Pyerin W, Burow E, Kinzel V (1983) Substrate-effected release of surface-located protein kinase from intact cells. Proc Natl Acad Sci 80:4021-4025
49. Volk HD, Josimovic-Alasevic O, Gross M, Diamantstein T (1989) The therapeutic efficacy of an IL-2 receptor monoclonal antibody correlates with an increase in serum soluble IL-2 receptor levels. Clin Exp Immunol 76:121-125
50. Honda M, Kitamura K, Matsuda K, Yokota Y et al. (1989) Soluble IL-2 receptor in AIDS. Correlation of its serum level with the classification of HIV-induced diseases and its characterization. J Immunol 412:4248-4255
51. Lang JM, Coumaros G, Levy S, Falkenrodt A et al. (1988) Elevated serum levels of soluble interleukin 2 receptors in HIV infection: correlation studies with markers of cell activation. Immunol Lett 19:99-102
52. Lambin P, Desjobert H, Debbia M, Fine JM, Muller JY (1986) Serum neopterin and $ß_2$-microglobulin in anti-HIV positive blood donors. Lancet 2:1216
53. Prince HE, Kleinman S, Williams AE (1988) Soluble IL-2 receptor levels in serum from blood donors seropositive for HIV. J Immunol 140:1139-1141
54. Franzitti F, Cavalli G, Foppa CU, Amprimo MC, Gaido P, Lazzarin A (1988) Raised serum $ß_2$-microglobulin levels in different stages of human immunodeficiency virus infection. J Clin Lab Immunol 27:133-137
55. Taylor DD, Black PH (1986) Shedding of plasma membrane fragments. Neoplastic and developmental importance. In: Steinberg MS (ed) Cell surface in development and cancer. Plenum, New York, pp 33-57
56. Emerson SG, Cone RE (1981) I-K^k and H-$2K^k$ antigens are shed as supramolecular particles in association with membrane lipids. J Immmunol 127:482-486
57. Doyle C, Shin J, Dunbrack RL, Strominger JL (1989) Mutational analysis of the structure and function of the CD4 protein. Immunol Rev 109:17-37
58. Rubin LA, Jay G, Nelson DL (1986) The released interleukin 2 receptor binds interleukin 2 efficiently. J Immunol 137:3841-3844
59. Semenzato G, Trentin L, Zambello R, Agostini C et al. (1988) Origin of the soluble interleukin-2 receptor in the serum of patients with hairy cell leukemia. Leukemia 2:788-792
60. Symons JA, Wood NC, Di Giovine FS, Duff GW (1988) Soluble IL-2 receptor in rheumatoid arthritis. Correlation with disease activity, IL-1 and IL-2 inhibition. J Immunol 141:2612-2618
61. Meuer SC, Hussey RE, Fabbi M, Fox D et al. (1984) An alternative pathway of T-cell activation: A functional role for the 50 kDa T11 sheep erythrocyte receptor protein. Cell 36:897-906
62. Sayre PH, Hussey RE, Chang H-C, Ciardelli TL, Reinherz EL (1989) Structural and binding analysis of A2 domain extracellular CD2 molecule. J Exp Med 169: 995-1009
63. Oh SK, Moolten FL (1981) Nonspecific immunosuppressive factors in malignant ascites: further characterization and possible relationship to erythrocyte receptors of human peripheral T cells. J Immunol 127:2300-2307

Structure and Function of Antigen Receptor Complexes on Murine B Lymphocytes

D. Haustein

Paul-Ehrlich-Institut, Paul Ehrlich Straße 51-59, D-6070 Langen, FRG

Introduction

B lymphocytes express immunoglobulin M (IgM) and D (IgD) on their surface that function as antigen receptors. When both Ig isotypes exist on one individual cell they possess the same antigen specificity. The binding of antigen to the receptors leads, together with other signals, to the activation of the cells. However, the mechanisms and structures by which Ig transmits such signals to the interior of the cells are unknown. One model of an antigen-recognition complex on the lymphocyte surface (Fig. 1) which is involved in the specific triggering of lymphocytes was postulated by us and others many years ago [5, 6]. By analogy with certain polypeptide hormones, this complex could consist of a recognition, a regulator and an effector element. Binding of antigen to the Ig molecule would induce its interaction with the regulator-effector complex which would then trigger a cascade of reactions that finally lead to the activation of the B cell. Since such receptor complexes have not yet been described, the aim of this work was to identify and biochemically characterize molecules which are associated and/or covalently linked to Ig. Furthermore, we are interested in the question whether IgM and IgD are linked to identical or different molecules. Various data have been reported indicating that IgM and IgD may have different functions [1,4,8,9,11,12]. Different anchor molecules might be responsible for the transmission of different signals induced by antigen binding to IgM and IgD.

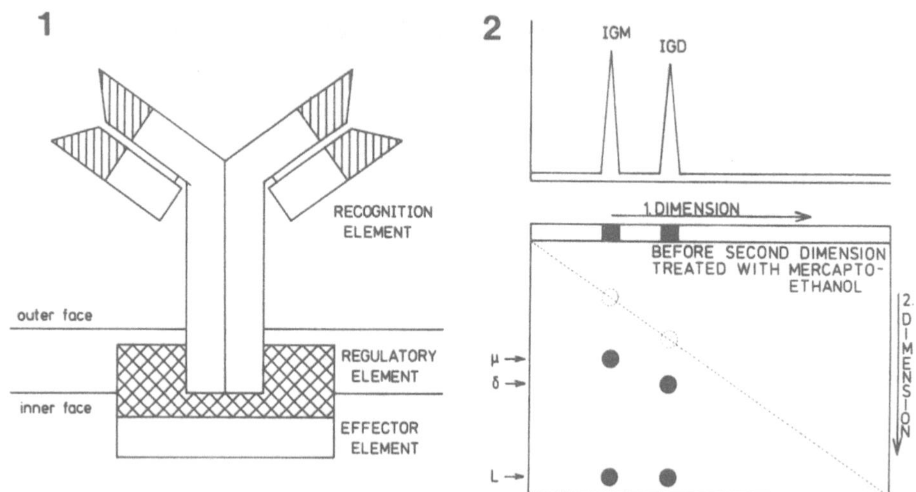

Fig. 1. A model of an antigen-receptor complex postulated by Marchalonis [6].
Fig. 2. Schematic diagram of 2D-SDS-PAGE (*on top*, separation of intact IgM and IgD by 1D-SDS-PAGE)

Results and Discussion

Surface proteins of viable murine B lymphocytes were radioiodinated by the lactoperoxidase method and subsequently solubilized by a nonionic detergent. Ig complexes were isolated by anti-Ig antibodies and heat-fixed protein-A-containing *Staphylococcus aureus* and then characterized by a very sensitive two-dimensional SDS polyacrylamide gel electrophoresis (2D-SDS-PAGE) which allows the identification of their subunit structure (Fig. 2). In the first dimension, intact (unreduced) Ig molecules were separated on an SDS gel. After cleavage of the disulfide bonds by mercaptoethanol, the obtained polypeptides were run on a second dimension SDS gel.

In the case of B cell surface Ig, four spots were expected: a μ chain, a δ chain, and two L chain spots (Fig. 2). However, a very complicated pattern was obtained. Figure 3 shows a schematic diagram of a pattern routinely obtained by 2D-SDS-PAGE of B cell surface Ig isolated by an L-chain-specific antiserum. There are various spots in the position of μ, δ and L

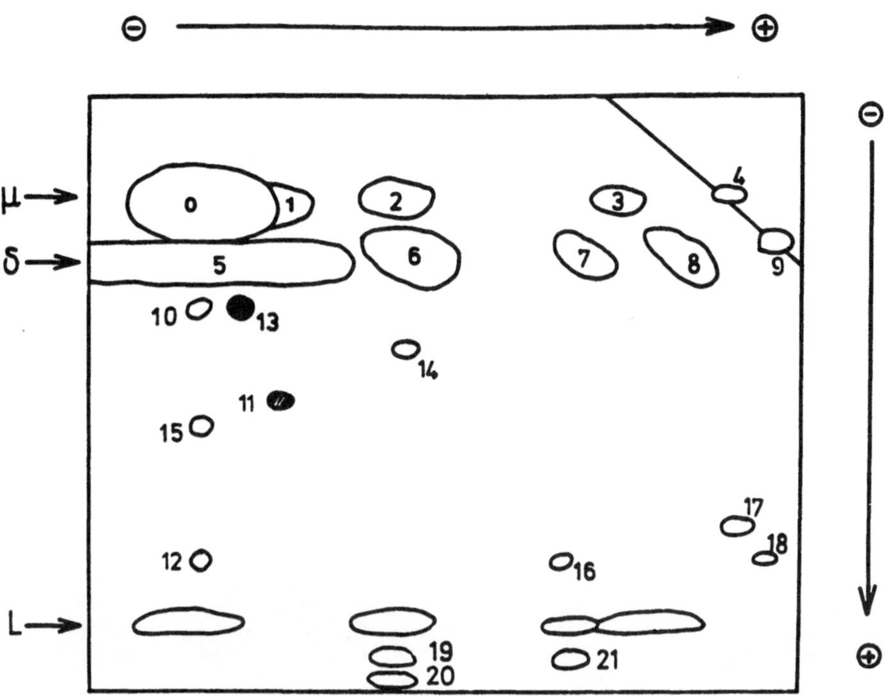

Fig. 3. Schematic diagram of ^{125}J-labeled Ig of B lymphocytes by 2D-SDS-PAGE (*first dimension*, unreduced, 6% acrylamide; *second dimension*, reduced, 8.5% acrylamide)

chain. In addition, there are spots on the diagonal which indicate noncleaved proteins. Various control experiments were performed in order to verify that this result was not an experimental artefact, but reflects the situation on the cell surface. The following control experiments are given as examples. Firstly, the addition of protease inhibitors (phenylmethane-sulfonylfluoride, *N*-ethylmaleimide, 1,10-phenanthroline) to the 1% Nonidet P4O used for the solubilization of the cells and to all buffers used subsequently did not influence the 2D-gel pattern of the isolated surface Ig molecules. Secondly, elution of IgM or IgD, incubation with a lysate of the nonradiolabeled spleen cells, and re-electrophoresis (Fig. 4) demonstrate that the spots of the 2D-gel shown in Fig. 3 represent Ig structures and Ig-linked proteins and that these structures are native B cell molecules and not degradation products.

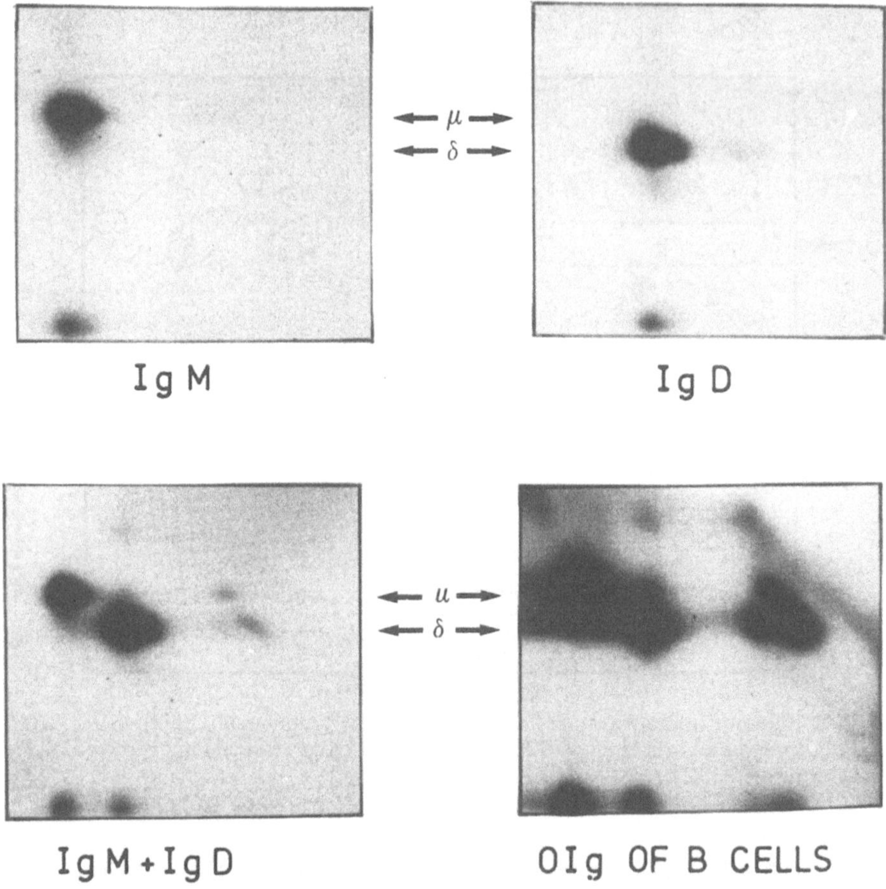

Fig. 4. 2D-SDS-PAGE of ^{125}J-labeled μ_2L_2 (IgM) and δ_2L_2 (IgD) after incubation with a 1% Nonidet P40 extract of radiolabeled spleen cells (*IgM + IgD*, μ_2L_2 and δ_2L_2 were pooled before incubation). For this, unreduced ^{125}J-labeled Ig of B lymphocytes isolated by an L-chain specific antiserum was separated by one-dimensional SDS-PAGE. μ_2L_2 and δ_2L_2 were eluted separately, incubated with spleen cell extract for 30 min and analyzed by 2D-SDS-PAGE (*OIg of B cells*, 2D-SDS-PAGE of B cell Ig isolated by an anti-L chain antiserum)

On the basis of the molecular weights of the intact structures in the first dimension and their identified polypeptides in the second dimension (using μ-chain and δ-chain specific antibodies), we have proposed various structures for Ig molecules

Table 1. Composition of Ig subunits of murine B lymphocytes

Spot no.	Molecular weight (KDa) 1D	2D	Heavy chain	Composition	Calculated molecular weight (KDa)
0	242	82	μ	μ_2L_2-S-S-(12) and/or	242
				μL-(10,12,15)	236
1	203	82	μ	μ_2L_2	212
2	150	82	μ	μL-S-S-(14)	156
3	100	82	μ	μL	106
5	200-350	62-69	δ	aggregates?	
6	140-150	62-69	δ	δ_2L_2 and/or	
				δL-S-S-(19,20)	15
7	100-110	62-69	δ	δL-S-S-(21)	109
8	85-90	62-69	δ	δL	93
9	65	69	-	albumin	

1D, one-dimensional; 2D, two-dimensional

and Ig complexes on the B cell surface [10]. These structures, including their molecular weights, are presented in Table 1. The main structure for surface IgM is an IgM half molecule to which proteins with molecular weights of 30 000, 45 000 and 55 000 are linked. Small amounts of μ2L2 and μL, disulfide-linked to a protein with a molecular weight of 50 000 and free μL are also present. The main IgD structure is an IgD half molecule to which two proteins with a molecular weight of 14 000 and two proteins with a molecular weight of 16 000 are linked. Furthermore, IgD half molecules disulfide-linked to a protein with a molecular weight of 16 000 and free half molecules can be observed.

As stated above, we are interested in the identification of proteins covalently as well as noncovalently linked to B cell surface Ig. After demonstrating the existence of S-S linked proteins, we asked whether additional proteins are noncovalently linked to membrane Ig. A technique to assess associations of membrane proteins is chemical crosslinking with bifunctional reagents [3]. For this, intact B cells were treated with different reagents and radioiodinated. We tested 16 different bifunctional reagents for their capacity to crosslink membrane Ig to other membrane proteins. Of these, only phosphorodiamidic anhydride (DPDA) was successful in detecting Ig-associated proteins.

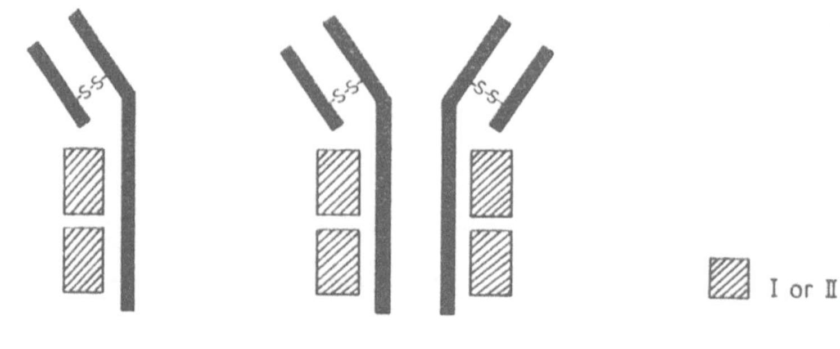

Fig. 5. Postulated Ig complexes. Association of IgM half molecules with protein I and II on murine B lymphocytes (*MW*, molecular weight)

When B cells were treated with DPDA and radioiodinated by the lactoperoxidase method, analysis of the isolated Ig by 2D-SDS-PAGE showed two Ig-associated proteins with molecular weights of 56 000 (protein I) and 46 000 (protein II). Both proteins are noncovalently linked to IgM-half molecules [5, 10]. On the basis of the molecular weights of the cleaved polypeptides in the second dimension, we proposed that there is a complex onB cells which consists of μL and two molecules of protein I and a complex of μL and two molecules of protein II (Fig. 5).

Proteins associated with IgD were not detected using DPDA. Either such proteins do not exist, or they are not radiolabeled by the lactoperoxidase method, or the crosslinking reagents used are not able to link them to IgD molecules. Table 2 shows the bifunctional reagents we applied to lymphocytes more recently. Various concentrations were tried in order to crosslink associated lymphocyte surface proteins. Using these reagents we did not detect IgD-associated proteins. On the other hand, these reagents were also not successful in crosslinking proteins I and II to IgM half molecules as demonstrated with DPDA. Treatment of B cells with these reagents resulted only in intra- and intermolecular crosslinking of the polypeptide chains of surface Ig.

Table 2. Crosslinking reagents used in this investigation in order to study the association (noncovalent linkage) of surface Ig with other membrane proteins on B lymphocytes

Crosslinking reagent	Formula	Cleaved by	Length (nm)
DPDA[1]	$BF_4^- N_2^+ \text{-} C_6H_4 \text{-S-S-} C_6H_4 \text{-} N_2^+ BF_4^-$	SH-reagents	1.5
BSOCOES[2]	(succinimidyl)$\text{N-O-C(=O)-O-}CH_2CH_2\text{-S(=O)}_2\text{-}CH_2CH_2\text{-O-C(=O)-O-N}$(succinimidyl)	base (pH 11.6)	1.3
EGS[3]	(succinimidyl)$\text{N-O-C(=O)-}CH_2CH_2\text{-C(=O)-O-}CH_2CH_2\text{-O-C(=O)-}CH_2CH_2\text{-C(=O)-O-N}$(succinimidyl)	NH_2OH	1.5
DTSP[4]	(succinimidyl)$\text{N-O-C(=O)-}CH_2CH_2\text{-S-S-}CH_2CH_2\text{-C(=O)-O-N}$(succinimidyl)	SH-reagents	1.2
SADP[5]	(succinimidyl)$\text{N-O-C(=O)-}CH_2CH_2\text{-S-S-}C_6H_4\text{-}N_3$	SH-reagents	1.1
DTPA[6]	$N_3\text{-}C_6H_4\text{-S-S-}C_6H_4\text{-}N_3$	SH-reagents	1.1

[1]4,4'-Diphenyldisulfide diazonium tetrafluoroborate; [2]Bis[2(succinimidooxy-carbonyloxy)ethyl]sulfone; [3]Ethylene glykol bis (succinimidylsuccinate; [4]Dithiobis(succinimidylpropionate); [5]Succinimidyl(4-azidophenyldithio)propionate; [6]Dithiobis(phenyl-4-azide)

With respect to our finding of a two-chain receptor molecule on B cells to which other proteins are linked, it is important to note that T cells also expose a two-chain receptor linked to a multichain complex, the T3 antigen. Anti-T3 antibodies can modulate T cell responses. In conclusion, antibodies against the Ig-linked proteins on B cells may be an extremely helpful tool to regulate antibody production.

As mentioned in the introduction, we postulate that Ig-linked proteins play an important role during the activation of B cells. They may, for example, transmit antigen-induced signals to the interior of the cells. In order to investigate this possibility, we use the stimulation of B cells by anti-Ig antibodies (Fig. 6). This model is extremely useful for our studies for many reasons: first, the anti-Ig antibody acts, like antigen, via the Ig receptors; second, B cells are stimulated polyclonally, and third, IgM and IgD can be triggered separately by anti-μ and anti-δ antibodies. The latter is important with regard to the question of whether different anchor molecules are responsible for the transmission of different signals (see Introduction). Figure 7 shows the result of a typical experiment in which B cells were stimulated with the monoclonal anti-μ antibody b-7-6 and polyclonal antibodies against IgM (MOPC 104E). As expected, rabbit anti-mouse IgG antibodies did not stimulate murine B cells. In this case, the proliferation rate was similar to that of B cells incubated with Ig (i.e., control). The proliferation rate of B cells stimulated with anti-δ antibodies (data not shown) was similar to that of B cells stimulated by the anti-μ antibodies b-7-6.

In order to study possible changes of the structure of Ig complexes during the activation, B lymphocytes were surface radioiodinated and subsequently stimulated by anti-μ or anti-δ antibodies. Ig complexes of stimulated cells were isolated and then analyzed by 2D-SDS-PAGE. Figure 8 shows a schematic diagram of a 2D-SDS-PAGE with Ig complexes of anti-μ, anti-δ and nonstimulated B cells. It is obvious that the pattern of Ig complexes of stimulated and nonstimulated cells are different. Furthermore, the Ig patterns of anti-μ and anti-δ treated cells are also distinct from each other. In addition, new spots can be detected in the Ig pattern of stimulated compared with nonstimulated cells, and there are spots which are more intense in the Ig pattern of cells treated with anti-δ than in the Ig pattern of those treated with anti-μ and nonstimulated cells.

Two explanations which are virtually exclusive can be given for these findings. First, structures which appear after the stimulation of B cells are radioioidinated non-Ig surface proteins which are linked to Ig after binding of anti-Ig (or antigen) to the receptor and are therefore likely to be involved in the subsequent steps of B cell activation. Second, the novel spots represent

Fig. 6. Stimulation of B lymphocytes by antigen and anti-Ig antibodies (*anti-μ* and *anti-δ*). **Fig. 7.** ^{3}H-thymidine incorporation into B lymphocytes induced by a monoclonal anti-μ antibody (b-7-6) or by a polyclonal goat anti-μ antibody (anti-MOPC 104E). Controls, goat IgG and rabbit IgG. **Fig. 8.** Schematic diagrams of 2D-SDS-PAGE of surface Ig of B lymphocytes. **A** Nonstimulated. **B** Stimulated by anti-μ for 1 h. **C** Stimulated by anti-δ for 1 h

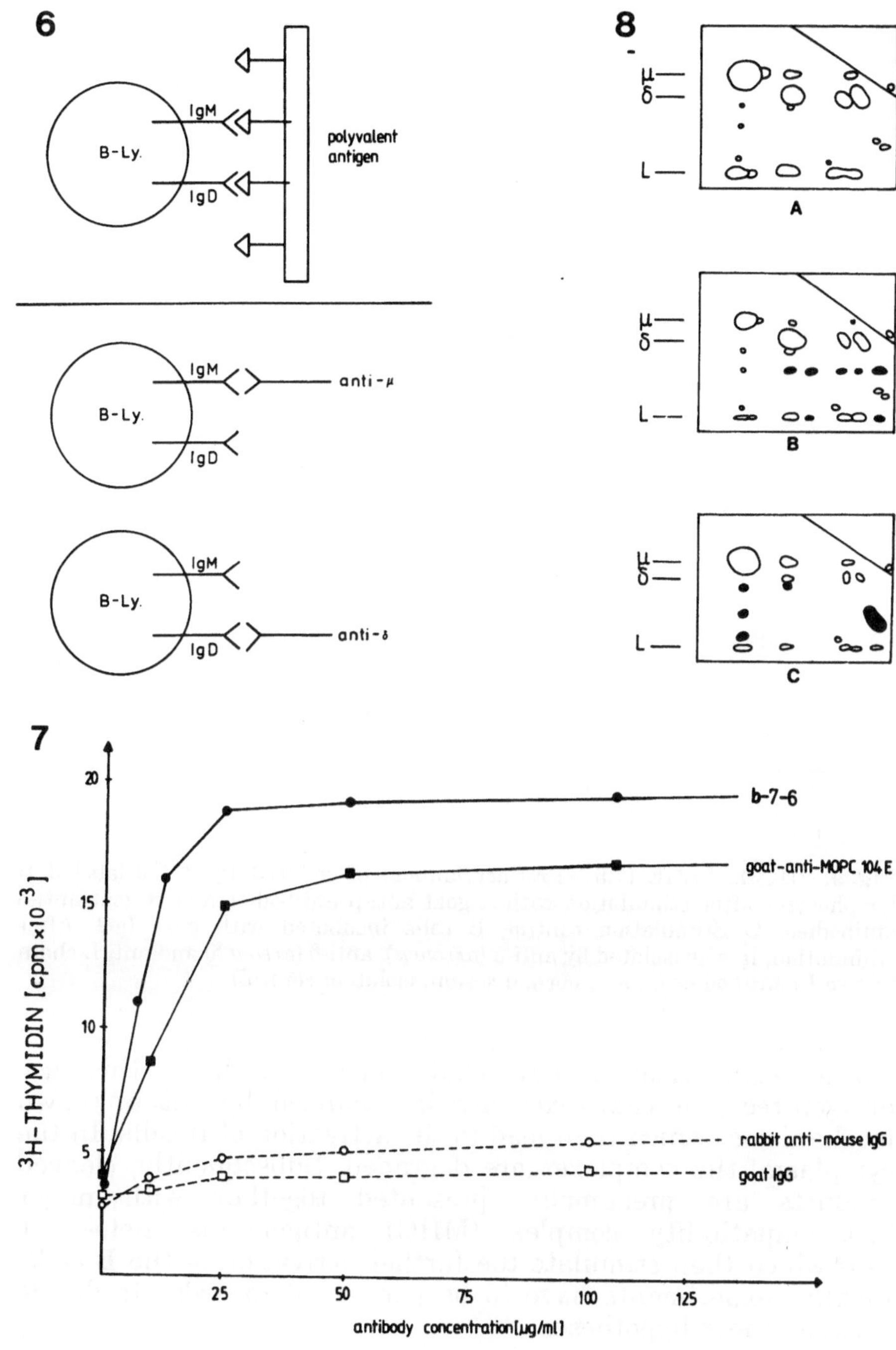
6
B-Ly.
IgM
IgD
polyvalent antigen
anti-μ
anti-δ
8
μ
δ
L
A
B
C
7
3H-THYMIDIN [cpm x10^-3]
20
15
10
5
25
50
75
100
125
antibody concentration [μg/ml]
b-7-6
goat-anti-MOPC 104 E
rabbit anti-mouse IgG
goat IgG

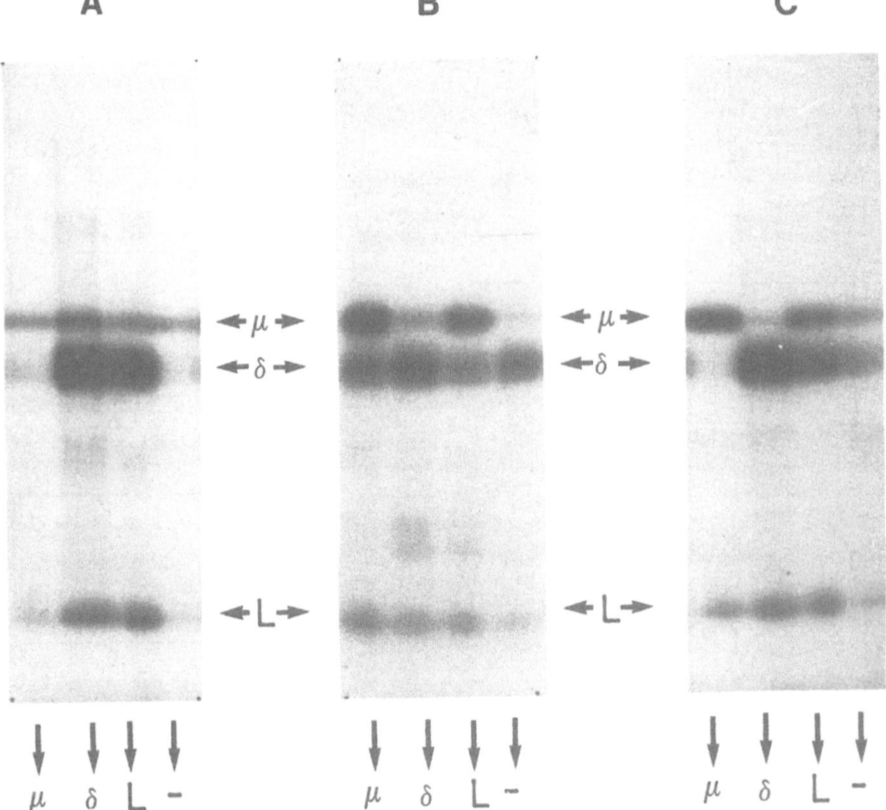

Fig. 9. 1D-SDS-PAGE (7.5%-15% acrylamide) of reduced Ig of [125]J-labeled B lymphocytes after stimulation with **A** goat anti-μ antibodies and **B** goat anti-δ antibodies. **C** Stimulation control, B cells incubated with goat IgG. After stimulation, Ig was isolated by anti-μ (*arrow μ*), anti-δ (*arrow* δ) and anti L-chain (*arrow L*) antibodies (*lane* -, normal serum, isolation control)

enzymatically cleaved heavy chains of surface Ig. In this case antigen-receptor complexes are internalized by yet unknown mechanisms, which then lead to the activation of B cells. In the cytoplasm, the complexes are degraded. Subsequently, cleaved products are presumably presented together with major histocompatibility complex (MHC) antigens to helper T cells,which then stimulate the further activation of the B cells. Further experiments have to be performed in order to decide between these hypotheses.

When Ig was isolated by μ-specific antibodies from surface radioiodinated, nonstimulated B lymphocytes and analyzed by a one-dimensional SDS-gel (Fig. 9C), only IgM and no IgD was observed as expected (Fig. 9C, lane μ). Likewise, anti-δ antibodies exclusively isolated IgD (Fig. 9C, lane δ), as anti-L-chain antibodies identified both IgM and IgD (Fig. 9C, lane L). However, in the case of B cells stimuated by anti μ antibodies (Fig. 9A), μ as well as δ chain specific antibodies isolated both isotypes. In particular, when Ig was isolated by anti-δ antibodies from anti-μ treated B cells (Fig. 9A, lane δ), a strong μ-chain band could be identified in addition to the expected δ-chain. The same phenomenon was observed for B cells stimulated by anti-δ antibodies: μ- and δ-specific antibodies isolated IgM as well as IgD (Fig. 9B).

These data clearly show that during the stimulation of B cells by anti-Ig antibodies both isotypes come in contact with each other in some way. This contact may be a direct interaction between both isotypes, resulting in a noncovalent linkage between IgM and IgD, or the Ig isotypes may use a connecting link, e.g., a protein, for this interaction.

The results presented here document for the first time an interaction between IgM and IgD on activated B lymphocytes. However, the biological function of this finding is not yet clear. One may speculate whether signals transmitted by IgM and IgD after binding of antigen are combined and then conferred to the interior of the cells by the same mechanism. In this connection it is interesting to mention the findings of Cambier and coworkers [2,7], who show that both Ig isotypes use the same mechanism to transmit their signals, namely the activation of protein kinase C.

Acknowledgments. This work was supported by grant Ha 793 from the Deutsche Forschungsgemeinschaft. I would like to thank Drs. L. Vogel and S.G. Norley for their critical reading of the manuscript.

References

1. Buck LB, Yuan D, Vitetta ES (1979) A dichotomy between the expression of IgD on B cells and its requirement for triggering such cells with two T-independent antigens. J Exp Med 149:987
2. Coggeshall KM, Cambier JC (1984) B cell activation. VIII. Membrane immunoglobulins transduce signals via activation of phosphatidylinositol hydrolysis. J Immunol 133:3384
3. Ji T (1983) Bifunctional reagents. Meth Enzymol 91:580
4. Kettmann JR, Cambier JC, Uhr JW, Ligler FS, Vitetta ES (1979) The role of receptor IgM and IgD in determining triggering and induction of tolerance in murine B cells. Immunol Rev 43:69
5. Koch N, Haustein D (1983) Association of surface IgM with two membrane proteins on murine B lymphocytes detected by chemical crosslinking. Mol Immunol 20:33
6. Marchalonis JJ (1976) Surface immunoglobulins of B and T lymphocytes: molecular properties, association with the cell membrane, and a unified model of antigen recognition. Contemp Top Mol Immunol 5:125
7. Monroe JG, Cambier JC (1983) B cell activation. I. Anti-immunoglobulin-induced receptor cross-linking results in a decrease in the plasma membrane potential of murine B lymphocytes. J Exp Med 157:2073
8. Sieckmann DG, Asofsky R, Mosier DE, Zitron IM, Paul WE (1978) Activation of mouse lymphocytes by anti-immunoglobulin. I. Parameters of the proliferative response. J Exp Med 147:814
9. Vitetta ES, Uhr JW (1977) IgD and B cell differentiation. Immunol Rev 37:50-88
10. Vogel L, Haustein D (1989) Immunoglobulin subunits of murine B lymphocytes: structure and association with other membrane proteins. Immunology 67:251
11. Weiner HL, Moorhead JW, Claman H (1976) Anti-immunoglobulin stimulation of murine lymphocytes. I. Age dependency of the proliferative response. J Immunol 116:1656
12. Zitron IM, Clevinger BL (1980) Regulation of murine B cells through surface immunoglobulin. I. Monoclonal anti-δ antibody that induces allotype-specific proliferation. J Exp Med 152:1135-1146

The Biology of Mammary Cancer

R. Dulbecco

The Salk Institute, P.O. Box 85800, San Diego, CA 92186-5800, USA

Introduction

My laboratory has been interested in the study of mammary cancer from cellular and molecular standpoints. Most of our work has been done using a model system in the rat, and I will here present an outline of this work. Mammary cancer can be easily induced in female rats, with characteristics similar to those of human cancers. To understand cancer development in this model, we studied the normal evolution of the cellular types during mammary gland development, and then determined the cellular types present in cancer at the various stages. We studied the mechanism of carcinogenesis, of progression to malignancy, and, finally, the role of genes in cancer development.

The mammary gland is made up of a system of ducts which branch out into the mammary fat pad. In the rat, the growing ducts are terminated by swellings called end-buds, which are responsible for duct elongation. Along the ducts, there are lateral buds which in pregnancy and lactation give rise to alveoli. The ducts are made up of two layers with cells of different lines: the luminal layer and the myoepithelial (basal) layer. The layering is not well recognizable in the end-buds. In all structures except the end-buds the cells are polarized, with a luminal surface provided with microvilli and a basal surface connected to the basal layer. The basal layer sits on the basal membrane, which is in turn surrounded by stroma which is probably highly specific.

Normal Cell Types

In all tissues, development derives from highly undifferentiated cells, the stem cells, which are capable of multiplication and differentiation. The evolution of cell types is well known in certain tissues such as the hematopoietic system, the epidermis, or the intestine. We studied the rat mammary gland in detail, because it was less well known, and the results obtained in the rat are generally applicable to humans. Our first approach involved the study of deoxyribonucleic acid (DNA) replication in cells present at the various sites in the gland in relation to the hormonal status of the animal. The second approach involved the use of specific markers.

DNA Replication

In the first approach [1], the hormonal status was determined by the phase in the normal ovarian cycle, which was identified by using vaginal smears. To detect DNA replication, tritium-labeled thymidine (H3TDR) was injected into the animal at specific phases of the cycle, causing the equivalent of about 2 h pulse. The glands were collected after either 1 h or a longer period (1 or 2 days) and labeled cells were detected in histological sections by autoradiography. In the glands collected after 1 h, the cells that undergo DNA replication at the time of injection are identified; in those collected later, the displacement of the cells in the subsequent period can be recognized. In these experiments the cells are in their normal environment and undergo normal developmental processes.

The main result of this approach is that a compartment of cells at the distal end or the end-buds are highly active when estrogen is elevated (proestrus or early estrus). Then, these cells move towards the lumen and the base of the end-bud. They are likely to be stem cells or their immediate product, especially in view of the fact that growth of the ducts into the fat pad is estrogen-dependent. In contrast, the duct cells themselves, especially those of the basal layer, are very active during late estrus or metestrus, when progesterone is elevated; the lateral buds and the alveoli, when present, are active at the same time.

Specific Markers

The second approach employs monoclonal antibodies as markers [2-4]. These were prepared using antigen suspensions of either tissue or cultures, which were inoculated into mice together with adjuvant. The monoclonal antibodies obtained were screened first against the immunizing antigen as well as unrelated antigens. Those showing adequate specificity were further studied by immunofluorescence for their localization on cryostat sections of rat mammary glands. Monoclonal antibodies to different types of keratins [5], to a surface protein present in basal cells [3], and to microvillin, a protein of microvilli turned out to be most useful in our studies [6]. The observations were usually made from double immunofluorescence, using a monoclonal antibody in conjunction with polyclonal antibodies to collagen IV, myosin, or laminin. A method was also developed for staining the same section with two different monoclonals [7].

The data were analyzed following the principle that, in a developmental tree, two cellular types sharing the same antigen are likely to be directly connected. This principle is illustrated by lymphocyte development [8]. Glands were studied beginning at the 16th day of fetal development when the early mammary ducts form, and then through all stages of development to lactation. The results [4] show that early ducts share the same markers as the basal layer of the epidermis from which they derive, i.e., basal keratin and a basal surface antigen, and in addition, have luminal keratin, which is the first mammary-specific marker.

Around the time of birth myosin-containing cells appear on the outside of the ducts at their distal ends. In sexually mature animals, the end-buds contain many layers of luminal cells containing luminal keratin, and a layer of basal cells containing basal keratin; the latter layer is either single, or if multiple, thin and irregular. The myosin-containing cells are at the same location as the cells activated in proestrus seen in the labeling studies, and are likely, therefore, to be stem cells. Microvillin appears in the center of the end-buds at places where small lumina form in the otherwise solid mass of cells, and it is also present at the lumen in all the ducts [9]. There is some evidence suggesting that in the ducts the basal cells give rise to luminal cells, based on the presence in some luminal cells of a keratin different from basal keratin, which is present in all basal cells, but not in most luminal cells [3, 4]. These special luminal cells maintain contacts with the basal membrane, suggesting a basal origin.

Discussion

These results indicate that, probably, there are two types of multipotent cells: the fetal cells present in the end-buds, which can give rise to the whole gland, and the basal cells of ducts, which can give rise only to ducts. This deduction was confirmed by transplantation experiments in mice (B. Armstrong and R Dulbecco, unpublished observations). In these animals, it is possible to isolate from a mammary gland, following some anatomical markers, a fragment containing essentially all the end-buds and some ducts, and another fragment containing only ducts. After collagenase digestion, the two types of fragments were injected into the fat pads of cycling female mice. After several weeks, all fragments gave rise to a proliferation of ducts, but only the end-bud containing fragments gave rise to both ducts and end-buds.

Mammary Gland Tumors

We can now consider the mammary gland tumors, which were studied using the same markers. Tumors were induced in 5-week-old rats by injection of *N*-nitrosomethylurea [10]. A high proportion of animals developed intraductal tumors, which, based on marker distribution, have the basic characteristic of either ducts or end-buds. There are, however, quantitative differences, such as multiple irregular luminal layers, and in-foldings of the wall in the tumors, which might be an exaggeration of the normal tendency of basal cells to migrate into the lumen [11]. The microvillin distribution is similar to that of end-buds. Because all cellular types are present in these tumors, they must derive from a multipotent cell, probably one of the fetal type, such as those present in the end-buds. This deduction is confirmed by the fact that the tumors arise, as discussed below, in estrogen-stimulated cells. An important characteristic is that the tumors are highly heterogeneous. That is, the marker distribution, although generally similar to that of normal ducts, can vary markedly from cell to cell. Often this heterogeneity gives rise to subclones with unusual distribution of markers. It seems, therefore, that there is in these tumors a variability which is both stochastic and clonal.

Oncogenes

It is known that rat mammary tumors of this kind contain an activated Ha-*ras* oncogene [12]. We have confirmed this observation for about 80% of the tumors we have studied (B. Armstrong and R. Dulbecco, unpublished). Which oncogenes are responsible for the remaining 20% is unknown. It has also been shown that the occurrence of these tumors is strongly dependent on the hormonal status of the animals. Thus, it has been shown that injection of the carcinogen into 2-day-old rats will give rise to a tumor only after sexual maturity is achieved. If maturity is prevented by ovariectomy, tumors do not appear, but they may appear later if the animals are given estrogen [13]. The possible significance of these observations is that tumors only appear when the proper cell type has developed, confirming our deduction that the tumors derive from estrogen-dependent stem cells present in the end-buds, or their immediate descendants. Cells at earlier stages of development, although carrying an activated oncogene, do not become neoplastic.

Tumor Development

The primary tumors are noninvasive, perhaps because they cannot be observed for a long time. In fact, they grow locally and rapidly ulcerate the skin. Their evolution, however, can be studied by serially transplanting them into isogenic rats [14]. We studied 12 transplant lineages, and in all of them we found that all tumors, sooner or later, become invasive; transplantation seems to hasten this process. In the various lineages, invasion does not develop in a uniform way: in some it occurs very early, in some others it occurs late and only in some sublines. Because the animals used were inbred and, therefore, uniform and the techniques always the same, this difference suggested that invasion is the result of a stochastic mutation-like event.

Further observations show that the development of the malignant properties is accompanied by stepwise changes in the organization of the tumors, a phenomenon known as progression [15]. The first change is the formation of structures that are in some way related to abnormal ducts. Out of these pseudoducts, cell strands emerge and invade the stroma. The strands are at first more than one cell thick, but later they are single-cell thick. Finally, isolated single tumor cells, identifiable for the presence of keratin, can be found in the stroma. In advanced stages, the cell markers are similar to those present in early fetal development. At that stage, the cells lack polarity, as shown by the coincidence of microvillin with laminin. The basal membrane persists during these changes and tends to disappear only in the very last stages. These results show that invasion results from progressive alterations of the genome accompanied by deregulation of the genes involved in normal development. Even more profound genetic changes are occasionally seen. Typical is the expression of the Thy-1 gene, which is not expressed during normal development, and, therefore, is likely to represent an ectopic gene expression [11]. Thy-1 expressing cells usually have no keratin, but contain abundant microvillin, which reveals their epithelial origin. They are the most malignant cells in animal transplantation.

Discussion

The interpretation of these data is that the early cell from which a tumor arises at first multiplies and differentiates along the normal pathway. Then the abilitiy to differentiate progressively disappears, until the tumor is made up of cells similar to the early cells. In this process epithelial characteristics such as polarity and production of the basal membrane are lost. Expression of keratin persists through many changes, as also observed in human breast tumors. The attainment of malignancy is, therefore, connected with the loss of ability to undergo differentiation.

The role of oncogenes in this process is not entirely clear. It seems likely that they are important at the beginning of the process, where a mutation in the Ha-*ras* oncogene causes a structural change in the protein, causing it to escape regulation. Less clear is the participation of oncogenes in progression, because in the study of the transplant lineages there is no relationship between the presence or absence of Ha-*ras* oncogene and the development of invasion (B. Armstrong and R. Dulbecco, unpublished). In fact, some lineages which initially had the oncogene lost it in subsequent passages, probably owing to unequal chromosome segregation; nevertheless they developed invasion. The Ha-*ras* oncogene, therefore, does not appear to be necessary for progression once it has started. An important observation is that a specific hormonal stimulation leading to the appearance of cells in a certain stage of differentiation, is required for tumor formation. This finding has important consequences for the mechanism of tumor latency, that is, the time lag between the initial event that gives rise to the tumor and the appearance of malignancy. It suggests that latency can be made up of two phases: one related to the differentiation of the cell in which the initiating lesion has occurred, to a stage suitable for tumor formation, and the other related to the time required for progression of the clone derived from such a cell from the initial changes to malignancy.

Conclusion

These observations made in rats have relevance for human breast cancer. In the first place, it is likely that human breast cancer is also initiated by the activation of an oncogene. However, it may be different from the oncogene found in the rat tumors, because only 5% of the human breast cancers show an activated H-*ras*. Another 5% show activated Ki-*ras*. The low proportion of cancers that are positive for these oncogenes can be attributed either to the activation of different unidentified oncogenes, or to the loss of the initiating oncogene during progression, as has been observed in some of the lineages of transplanted rat tumors. It would not be surprising if the oncogenes responsible for differentiation were different in rats and humans because of the species difference, and also because the dosage of the carcinogenic agent would be quite different, being much smaller in women. The phenomenon of progression may occur along very similar lines in the two species in terms of cellular changes. In women, we know more about the genetic changes that accompany progression. Much attention has been given to the amplification of oncogenes such as *neu*, although the results are at the present time somewhat controversial. Considerable attention is now being given to the state of anti-oncogenes; loss or alteration of p53 and retinoblastoma expression is frequently observed in human breast cancers, and it is likely that it contributes to progression.

Another point of interest is that gene amplification is very extensive in human cancers together with other chromosomal alterations such as translocations and aneuploidy. It seems most likely that these are consequences of the destabilization of the genome caused by the initial activation of the oncogene. They may be due to alteration of the normal expression of the proteins involved in chromosome segregation, recombination, and amplification. Another consequence of these studies has to do with the meaning of latency, which for human breast cancer can be as long as 30-40 years. This long latency is seen from the influence of events occurring in youth, such as the age of menarche, age, and number of pregnancies, on future cancer development. The studies in the rat suggest, as I have already pointed out, two types of latency: one related to cell differentiation, and the other to progression [16]. A latency related to cell differentiation may also play a role in women, because the conditions predisposing to breast cancer involve hormonal events capable of affecting stem cell development. However, we know very little about the latter point. If the time period before the necessary differentiation represents an important component of cancer latency, the possibilities for early diagnosis would be diminished because the time for the development of progression would become shorter. Perhaps such a shortening could be related to the relatively modest effect of yearly mammography on mortality from breast cancer. Perhaps more frequent mammograms would be advisable.

Acknowledgments. This investigation was supported by Grant 1-RO1 CA21993 from the National Cancer Institute and by grants from The Hammer Foundation, the Joseph Drown Foundation, the Elsa U. Pardee Foundation, the Max and Victoria Dreyfus Foundation, and the Rice Foundation. The research conducted in part by the Clayton Foundation, California Division. R. Dulbecco is a Senior Clayton Foundation Investigator.

References

1. Dulbecco R, Henahan M, Armstrong B (1982) Cell types and morphogenesis in the mammary gland. Proc Natl Acad Sci USA 79:7346-7350
2. Dulbecco R (1982) Immunological markers in the study of development and oncogenesis in the rat mammary gland. J Cell Physiol Suppl 2:19-22
3. Dulbecco R, Unger M, Armstrong B, Bowman M, Syka P (1983) Epithelial cell types and their evolution in the rat mammary gland determined by immunological markers. Proc Natl Acad Sci USA 80:1033-1037
4. Dulbecco R, Allen WR, Bologna M, Bowman M (1986) Marker evolution during the development of the rat mammary gland: stem cells identified by markers and the role of myoepithelial cells. Cancer Res 46:2449-2456
5. Allen R, Dulbecco R, Syka P, Bowman M, Armstrong B (1984) Developmental regulation of cytokeratins in cells of the rat mammary gland studied with monoclonal antibodies. Proc Natl Acad Sci USA 81:1203-1207
6. Allen R, Dulbecco R, Syka P, Bowman M (1984) Microvillin: a 200-kilodalton protein in microvilli of rat mammary cells detected by a monoclonal antibody. Proc Natl Acad Sci USA 81:5459-5463
7. Bologna M, Allen R, Dulbecco R (1986) A method for double immunofluorescent staining by the indirect procedure with antibodies of the same isotype. J Immunol Methods 86:151-153
8. McKenzie IFC, Potter T (1979) Murine lymphocyte surface antigens. Adv Immunol 27:179-321
9. Dulbecco R, Allen WR, Bowman M (1984) Lumen formation and redistribution of inframembranous proteins during differentiation of ducts in the rat mammary gland. Proc Natl Acad Sci USA 81:5763-5766
10. Gullino PM, Pettigrew HM, Grantham FH (1975) N-nitrosomethylurea as mammary gland carcinogen in rats. J Natl Cancer Inst 54:401-414
11. Dulbecco R, Armstrong B, Allen WR, Bowman M (1986) Distribution of developmental markers in rat mammary tumors induced by *N*-nitrosomethylurea. Cancer Res 46:5144-5152
12. Zarbl H, Sukumar S, Arthur AV, Martin-Zanca D, Barbacid M (1985) Direct mutagenesis of Ha-*ras*-1 oncogenes by *N*-nitroso-*N*-methylurea during initiation of mammary carcinogenesis in rats. Nature 315:382-385
13. Kumar R, Sukumar S, Barbacid M (1991) Activation of *ras* oncogenes precedes the onset of neoplasia. Science 248:1101-1104
14. Dulbecco R, Armstrong B (1988) Stochastic development of invasive potential in rat mammary tumors induced by *N*-methyl-*N*-nitrosourea. Proc Natl Acad Sci USA 85:8659-8663
15. Dulbecco R, Armstrong B, Allen R (1988) Reversion toward an earlier stage of differentiation and loss of polarity during progression of *N*-methyl-*N*-nitrosourea-induced rat mammary tumors. Proc Natl Acad Sci USA 85:9292-9296
16. Dulbecco R (1989) Experimental studies in mammary development and cancer: relevance for human cancer. Adv Oncol 5:3, 3-6

Virus Quasispecies

M. Eigen

Max-Planck-Institut für biophysikalische Chemie,
Göttingen, FRG

The "quasispecies" concept originates from the theory of mutating, self-replicating systems. It accounts for the mutant distribution of a wildtype sequence, which as a whole is subject to selective evaluation. The stability of such a distribution depends on the mutation rate, the length of the reproducing nucleic acid sequences, and the selective values of individual mutants or their mean. There is a threshold value of mutation rate (error threshold) that must not be surpassed; otherwise, the quasispecies deteriorates by an accumulation of mistakes (error catastrophe). The disintegration of the quasispecies (centered around a wildtype) following a violation of its error threshold or its delocalization to a new wildtype has the character of a phase transition; physically the system is similar to a spin glass. The optimal conditions for evolutionary adaptation are just below the error threshold. Here the quasispecies consists of a mutant spectrum in which the wildtype (i.e., the "center of gravity") is present in low amounts relative to the total number of mutants. Mutants which resemble most closely the selective value of the wildtype appear with highest abundance. As a consequence of the selected distribution of mutants along ridges and peaks of the value landscape, evolutionary adaptation proceeds with optimal speed along some internally guided routes.

The quasispecies can be represented as a localized distribution in sequence space, a point space where each point is attributed to one of the 4^{ν} possible sequences of length ν and where all points are arranged to correctly represent sequence kinships. In other words, any pair of mutants differing in k positions must be separated by a distance k. Since nucleic acids make use of four symbol classes (nucleotides), sequences of length ν require a 2ν-dimensional space.

The quasispecies model could be tested quantitatively with single stranded ribonucleic acid (RNA) viruses which are the simplest "natural" representatives of living systems. Available are sequence data for three different types of viruses:

- Plus-strand viruses such as coliphage Qß and its variants, polio-viruses and picorna viruses
- Minus-strand viruses such as influenza A viruses and vesicular stomatitis (VS) viruses
- Retroviruses such as the acquired immunodeficiency syndrome viruses human immunodeficiency virus (HIV)-1 and HIV-2, and various simian viruses such as simian immunodeficiency virus (SIV)

The RNA sequences of these viruses contain between approximately 5 000 and 15 000 nucleotides and are replicated at mutation rates which are many orders of magnitudes larger than those known for autonomous microorganismus, up to 10^{-5}-10^{-4} substitutions per nucleotide per replication round. The replication of these viruses near their error threshold enables them to adapt very quickly to changing environments. The extent to which they use this flexibility to evade the immune response of an infected organism depends also on the fitness landscape of the mutant spectrum (in particular on the number of neutral or nearly neutral mutants that can form before immunity is established).

Comparative analysis of sequence data has been performed using a new method called statistical geometry in sequence space. In this method, quartets of sequences are projected into the multidimensional sequence space, and geometries are established from the nodes of all connecting lines. The method becomes a statistical one by averaging the data of all $\binom{n}{4}$ possible quartets in a set of n sequences. The geometries obtained include up to eight classes of distance segments that form figures with mixed box-like and tetrahedral structures. The various segments refer to defined classes of forward, parallel, and reverse mutations and allow an assessment of kinship topologies and their degree of randomization. Considering the mutual relations among the various distance segments provides an internal calibration and allows an establishment of kinship relations among sequences in a more precise way than is possible by comparing overall distances only and fitting them into dendrograms.

In this analysis influenza-A virus appears as a narrowly confined quasispecies with well defined wildtype that varies uniformly in a relatively short time, while HIV shows a comparatively larger scattering of the quasispecies structure due to parallel (nearly) neutral mutations. The older nodes among HIV and SIV sequences date back several hundred years and show a large degree of randomization. The earliest nodes of HIV that were isolated in the Western world (Europe and USA) are not more than 20 years old. Polio viruses (against which we have now effective immune protection) differ in their topological appearance from both influenza-A and HIV due to the much more symmetrically extended tree-like structures of their evolving wildtype.

The quasispecies nature of virus populations is responsible for their flexibility towards environmental changes. These effects can be quantitatively studied in cell-stat evolution reactors. The cell-stat is a flow reactor that is fed with a constant flux of cell culture and that allows a study of virus infection in both transient and steady state under controlled environmental conditions. Error rate and selection pressure are the forces that drive the evolutionary changes to be observed at the genotypic and phenotypic level. Experiments of this kind not only allow a closer study of the defense mechanisms of viruses against immune attack, but also may serve as model scenarios for the early molecular evolution of life.

Retroviruses: Nature's Genetic Engineers

M. Temin

McArdle Laboratory, 1400 University Avenue, University of Wisconsin, Madison, Wisconsin 53706, USA

I am going to describe a family of viruses that have an incredible ability to alter themselves and to alter organisms. These viruses are called retroviruses because of their ability to go back and forth between the external world and the chromosome, i.e., ribonucleic acid (RNA) in the external world and deoxyribonucleic acid (DNA) inside the cell. Furthermore, these viruses use error-prone processes, and so they have evolved into many different forms and varieties and have helped to shape the nature of eukaryotic genomes. Viruses are shaped by the forces of evolution, not by history. Viruses affect humans in ways shaped by human behavior, which is reflected in history as epidemics [1]. As emphasized by Eigen (this volume), viruses exist as populations or swarms with each individual unique [2]. However, this description also applies to most populations of organisms, including the human population.

Several per cent of human, mouse, and *Drosophila* genomic DNA has been inserted from outside. Certain cancers in animals, including some that occur outside laboratories, are caused by the insertion of modified genes into cell DNA. Acquired immunodeficiency syndrome (AIDS) results from a virus adding its genome to human T-cell DNA. Tumor cells from a human cancer patient have recently been marked by a virus and reinjected into the patient, a precursor of gene therapy in humans. Retroviruses are a common feature of all of these phenomena. Viruses are intracellular parasites with a simple genome structure which occur in many forms. Retroviruses differ from most other viruses by having both DNA and RNA genomes at different times in their life cycle [3, 4]. However, having both types of genomes requires retroviruses to go through some molecular gymnastics to recreate a 5' promoter to synthesize the RNA genome and to synthesize a copy of the primers for synthesis of the DNA genome. In addition, retroviruses have an efficient means of integrating their DNA genome into the cell

DNA. These features make retroviruses genetic engineers of cells, because they add genes to the cell genome and do not necessarily kill their host cell by replication, using instead the cellular transcription machinery to synthesize their RNA genomes.

The retrovirus genome is spatially separated into terminal regions with *cis*-acting sequences and internal portions with protein-coding or *trans*-acting sequences. Thus, retroviruses are easily modified by nature and by scientists. For example, highly oncogenic retroviruses have their protein-coding sequences altered so that the retrovirus expresses an altered cellular protein related to growth control and causes cancer in infected animals. In the laboratory, analogs of highly oncogenic retroviruses have been constructed with other inserted genes. These modified retroviruses are called retrovirus vectors [5]. Since most retrovirus vectors are defective, helper, or packaging cells have been constructed to supply the missing retrovirus protein [6]. Helper cells transfected or infected with retrovirus vectors produce helper virus-free retrovirus vector virions. Helper virus-free retrovirus vectors have been used for the expression of proteins like oncogenes and receptors, for genetic engineering of farm and laboratory animals, for human gene therapy, and for the study of retrovirus replication.

My laboratory has used retrovirus vectors to study retrovirus variation, the raw material of retrovirus evolution. For this study, we devised a protocol that isolates a single cycle of retrovirus replication (Fig. 1) [7]. We also devised a shuttle vector which replicates as a retrovirus in mammalian cells and as a plasmid in bacterial cells [8]. The vector also contains a *lacZ*α peptide gene as a reporter of spontaneous mutations.

There is a high base pair (bp) substitution rate in retrovirus replication. But there are also high rates of frameshift mutations, deletions, and deletions with insertions. In addition, one in ten thousand proviruses studied contained a 1 kbp region with a 2% bp substitution rate. We call such a high rate of mutation hypermutation, and we hypothesized that hypermutation results from an epigenetic change in a polymerase. Furthermore, there are hot spots for frameshifts and deletions. Tandem repeats disappeared at a rate of 40% per replication cycle, and frameshifts occurred in runs of 10 T at the same high rate. One complex deletion with insertion removed the primer for plus-strand DNA synthesis and so must have occurred during plus-strand DNA synthesis. The sequence of this and other deletions

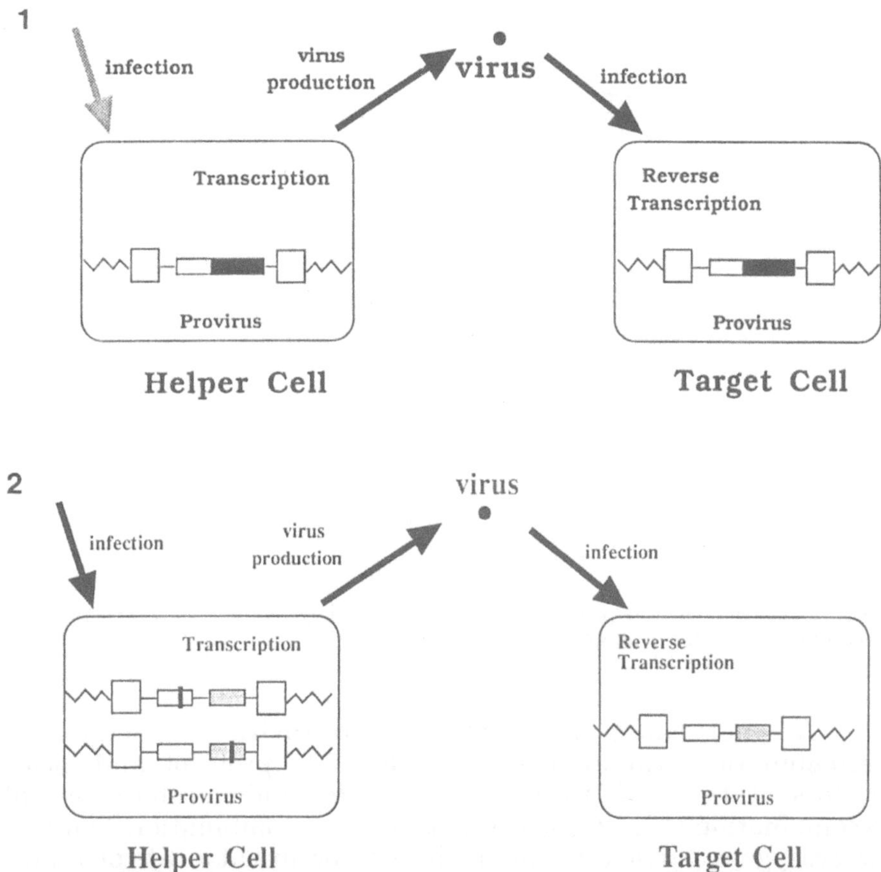

Fig. 1. Single cycle of retrovirus replication. The single cycle starts with a provirus in the helper cell *on the left* and ends with the provirus in the target cell *on the right. Open boxes* are long terminal repeats; *open and filled rectangles* are inserted genes; and *zig-zag lines* are cellular DNA. **Fig. 2.** Single cycle of retrovirus replication and vectors to study recombination. (Symbols are as described in the legend of Fig. 1.) These vectors express two genes. The *vertical bars* mark mutations in the inserted genes. The recombinant provirus in the target cell expresses both inserted genes

with insertions demonstrated that they resulted from a shift of the growing point during either minus- or plus-strand DNA synthesis to an improper template formed from a portion of the parental virus RNA.

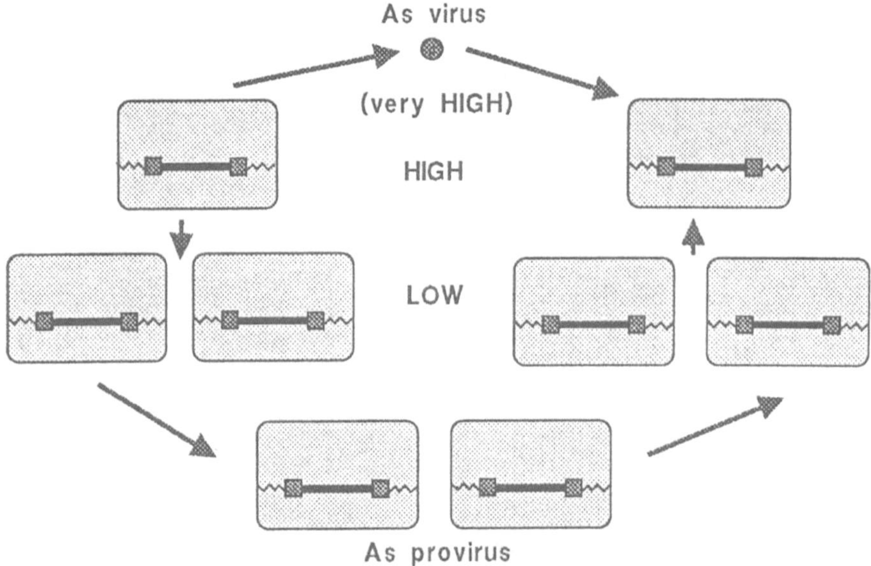

Fig. 3. Alternate modes of retrovirus replication (key as for Fig. 2, except filled rectangles are retrovirus genes)

Another cause of variation in retrovirus replication is recombination. An analogous single cycle protocol and other vectors were used to study the rate and mechanisms of recombination (Fig. 2) [9, 10]. Retrovirus recombination requires heterozygous particles, is not reciprocal, occurs at a rate of 4% for markers 1 kbp apart, and results in the formation of only a single provirus. Using vectors with eight different restriction enzyme cleavage site markers, we found several different patterns of recombination. About half of the proviruses had more than one template switch. Furthermore, use of the markers in the long terminal repeats revealed a correlation between interstrand minus-strand strong stop DNA transfer and a single internal template switch, and between intrastrand minus-strand strong stop DNA transfer and more than one internal template switch. Further studies indicate that all recombination takes place during minus-strand DNA synthesis and that intrastrand minus-strand strong stop DNA transfer occurs when there are breaks in viral DNA (Hu and Temin, to be published).

Retrovirus replication can be very error prone when hypermutation occurs, error prone when bp substitutions, frameshifts, deletions, and deletions with insertions occur during normal retrovirus replication, and less error prone when replication occurs as a provirus with cellular repair process (Fig. 3). In addition, the high rate of recombination shuffles variant genomes to make an even larger pool of variants. These alternate modes of retrovirus replication, with high variability in one mode to generate diversity and low variability in another mode to fix successful variants, help to explain their evolutionary success. This evolutionary success is seen in the proliferation of numerous different types of retroviruses, including human immunodeficiency virus (HIV)-1, and in the difficulty of developing a successful vaccine against HIV-1 and AIDS. In addition, this evolutionary malleability allows the use of retroviruses for genetic engineering in the laboratory, in animal husbandry, and in human medicine. Retroviruses help to remind us that we are not the only living creatures on this planet, as this dedication reminds us that we must not forget history.

Acknowledgments. The research in my laboratory is supported by Public Health Service grants CA-22443 and CA-07175 from the National Cancer Institute. I am an American Cancer Society Research Professor.

References

1. McNeil WH (1976) Plagues and peoples. Anchor, New York
2. Eigen M, Biebricher CK (1988) Sequence space and quasispecies distribution. In: Domingo E, Holland JJ, Ahlquist P (eds) RNA virus genetics, vol 3. CRC Press, Boca Raton, pp 211-245
3. Weiss R, Teich N, Varmus H, Coffin J (eds) (1982) Molecular biology of tumor viruses: RNA tumor viruses, vol 1, 2nd edn. Cold Spring Harbor, NY
4. Weiss R, Teich N, Varmus H, Coffin J (eds) (1985) RNA Tumor viruses: Molecular biology of tumor viruses: RNA tumor viruses, vol. 2, supplements and appendixes. Cold Spring Harbor, NY
5. Temin HM (1989) Retrovirus vectors: promise and reality. Science 246:983
6. Miller AD (1990) Retrovirus packaging cells. Hum Gene Ther 1:2-14
7. Dougherty JP, Temin HM (1986) High mutation rate of a spleen necrosis virus-based retrovirus vector. Mol Cell Biol 6:4387-4395
8. Pathak V, Temin HM (1990) Broad spectrum of in vivo forward mutations, hypermutations, and mutational hotspots in a retroviral shuttle vector after a single replication cycle. I. Substitutions, frameshifts, and hypermutations.

II. Deletions and deletions with insertions. Proc Natl Acad Sci USA 87:6019-6023 and 6024-6028

9. Hu W-S, Temin HM (1990) Retrovirus recombination and reverse transcription. Science 250:1227-1233

10. Hu W-S, Temin HM (1990) Genetic consequences of packaging two RNA genomes in one retroviral particle: pseudodiploidy and high rate of genetic recombination. Proc Natl Acad Sci USA 87:1556-1560

Retroviruses and Autoimmunity [1]

J.R. Kalden

Department of Internal Medicine III and Institute for Clinical Immunology and Rheumatology, University of Erlangen-Nürnberg, Krankenhausstraße 12, D-8520 Erlangen, FRG

Retroviruses have been discussed for years as an aetiological agent in autoimmune and different rheumatic diseases. However, despite intensive research activities, the role of retroviruses in the pathogenesis of either autoimmune or rheumatic diseases still remains unproven. The observation that arthrocutaneous disorders including Reiters's syndrome, psoriasiform rashes and other forms of chronic arthritis occur with an increased prevalence in human immunodeficiency virus (HIV)-1-infected persons [1, 2] has reactivated the discussion of retroviruses as aetiologically important agents in the pathogenesis of autoimmune diseases. A minority of patients with HIV infections might present autoantibodies including anticytoplasmic antibodies [3, 4] and antibodies against the lupus anticoagulant [5]. However, these antibodies are of low titre and probably reflect a polyclonal B-cell activation known to occur in HIV-infected persons. Finally, an association of HIV-1 infections with the appearance of Sjögren's syndrome [6] and vasculitis [7] has been reported.

Retroviruses are RNA viruses which by definition replicate through a DNA intermediary [8]. Three subfamilies of retroviruses are known: RNA tumour viruses or oncoviruses, slow viruses or lentiviruses, and foamy or spuma viruses. Murine lupus type C infectious oncoviruses have intensively been studied as being involved in the aetiology of murine systemic lupus erythematosus (SLE). As shown by Steinberg and coworkers [9, 10] only recently, a certain subclass of type C retroviruses, the mink cell focus forming virus (MCF), might indeed be involved in the aetiopathogenesis of SLE in lupus prone mouse strains.

In the context of this communication, emphasis will be put on the role of lentiviruses in rheumatic diseases. The most

[1]Part of the work was supported by Deutsche Forschungsgemeinschaft

remarkable characteristics of lentivirus infections are shown in Table 1. There is an obviously long variable incubation time, and in addition to the virus infection, cofactors for the establishment and perpetuation of the infection seem to be necessary. Thus, as shown by Banks et al. [11] on the caprine arthritis-encephalitis animal model, severe progressive arthritis was predominantly present in infected animals in joints which were additionally inoculated with methylated human serum albumin. This severe progressive arthritis with an excessive macrophage and T-cell proliferation was not shown in infected animals which were intra-articularly inoculated with normal saline. From these experiments it seems that activated macrophages costimulated by a third party antigen appear in the synovial membrane and fluid and lead to severe progressive arthritis by the release of cytokines. Working on the bovine lentiviral disease visna, Kennedy-Stoskopf [12] showed a great similarity between this retrovirally induced disease and human rheumatoid arthritis (RA). With regard to other characteristics of lentiviral infections, only a certain percentage of infected persons will develop fullblown diseases, which accounts for additional genetic factors, and finally, it is well known that retrovirus infections might cause secondary infections by the impairment of the immune system, the cells of which are the main target of lentiviruses.

Interactions Between Retroviruses and the Immune System

Before addressing in more detail the possible role of retroviruses in human rheumatic diseases, some comments seem to be necessary regarding the interaction of retroviruses with the immune system (Table 1). Both infectious and endogenous retroviruses can modify immune reactivity by two major mechanisms. A variety of infectious onco- and lentiviruses can cause immunosuppression. Thus, for the HIV-1 envelope protein gp120, it has been demonstrated that it inhibits antigen-induced peripheral lymphocyte proliferation [13], which could also be shown for a transmembrane envelope retroviral protein p15E. p15E inhibits feline, murine and human lymphocyte proliferation

Table 1. Characteristics of lentivirus infections

Long variable incubation time
Involvement of cofactors
Only a certain percentage of infected persons develop full-blown disease: genetic factors
Causative for secondary infections by the impairment of the immune system
Infectious for cells of the mononuclear system
Modification of immune function
- Direct immunological effects: immunosuppression by retroviral proteins (e.g. p15E)
- Effects on the host immune response

[14, 15] as well as human T-cell IL-2 secretion [16]. The mechanisms of different in vitro effects of the protein p15E or homologous peptides, respectively, are not fully understood, and further studies seem to be necessary to gain further insight into the immunomodulary mechanisms as induced by p15E. Similarly, a homologous peptide to the transmembrane envelope protein of HIV-1, the gp41, has also been discussed as having immunosuppressive effects similar to those of the synthetic analogue of the p15E [17].

Besides these direct effects of infectious onco- and lentiviruses on the immune system, indirect immune effects on the host response to retroviruses include the release of interferon-α. Interferon-α can induce MHC class II expression on antigen-presenting cells and by expressing class II antigens; even autoantigens might be presented to T-cells, resulting in autoimmune phenomena, as recently discussed by Feldmann et al. [18]. Furthermore, as already proposed in 1983 by Plotz [19] the formation of antibodies to viral components that bind to cell surface molecules might serve as antigens for anti-idiotypic antibodies. Anti-idiotypic antibodies could then act as autoantibodies against the cell structure to which the virus is directed. Finally, since retroviral proteins might share amino acid homology to host proteins [20], antibodies or cytotoxic lymphocytes directed at a certain retroviral protein might cross-react with homologous autologous structures and could thereby induce autoimmunity. For example, the HIV gp41 protein shares a 5 amino acid sequence with a conserved region of class II MHC proteins and by mechanisms called molecular mimicry, autoantibodies might be produced that cross-react with the class

Table 2. Circumstantial evidence for the possible involvement of retroviruses in the aetiopathogenesis of SLE

Demonstration of atypical interferon in SLE sera
Demonstration of reverse transcriptase activity in culture supernatants of SLE patients
Demonstration of HIV-1 homologous DNA from SLE plasma
Demonstration of cross-reactivity between anti-68kDa-snRNP and retroviral gag protein

Analysis of Nucleic Acids from SLE Patient's Plasma

SLE, a prototype of an autoimmune disease in humans, is, besides its clinical features, characterized by the appearance of different autoantibodies, including those against double-stranded human B-DNA, which are not only disease-specific but also discussed as being of major importance for SLE pathogenesis.

Although from recent animal studies as well as from the human disease our knowledge has considerably increased with regard to the nature of anti-dsDNA antibodies, the target antigen and the triggering mechanisms for autoantibody formation against B-DNA remain unsolved. To obtain further insight into trigger mechanisms of anti-dsDNA antibodies and to identify the possible autoantigen, immune complexes as isolated from plasmapheresis fluid from patients suffering from active SLE were analysed. Applying different methods, the antibodies were separated from the antigen of the immune complexes, and the antigen was furthermore subjected to various analytical steps. Biochemical analysis of the plasma nucleic acids revealed a high molecular DNA fraction of approximately 20 kbp and an RNAse insensitive RNA of more than 60 b. The isolated nucleic acids proved to be highly immunogenic, including the formation of monoclonal antibodies to natural DNA as well as to synthetic polynucleotides. A detailed biochemical and molecular cloning analysis of the immunogenic DNA has been previously published [28, 38]. For further analysis, 13 plasma nucleic acids were cloned and sequenced. Using a suitable DNA analysis program, various sequence homologies were detected. The most striking one was one clone (E6) showing a homology of more than 80% with the gag-pol overlap region of the human immunodeficiency virus HIV-1. This clone was further analysed [34, 35], however, when applying E6-derived flanking primers for amplification by means of polymerase chain reaction (PCR), short retroviral DNA sequences were not only detected in 7 out of 13 investigated SLE patients' peripheral blood lymphocytes, but also in lymphocytes of patients with other autoimmune diseases as well as in normal controls. Also using E6 as a probe, Krieg and Steinberg (personal communication) found on Southern blots good hybridization at moderate stringency to multiple nonpolymorphic bands in all of the patients and controls tested. These data from Southern blot analysis as well as of the PCR suggest that the clone E6 could be derived from, or is, a homologous retroviral element. If it can be proven by further experiments that the cloned plasma DNA from SLE patients with the described 81.6% homology over an overlap of 174 b with the pol-gag region of HIV-1 was part of an endogenous retrovirus, this could still have some implication for the aetiopathogenesis of SLE. Our own reported data that messenger RNA with a sequence homology to the E6 clone could only be demonstrated in EBV-immortalized B-cell clones from SLE patients, but not from controls, imply that a protein derived from this messenger RNA might be produced and expressed. In this context it is of interest that Krieg et al. [36] recently showed that a transmembrane protein (formerly called p15E) of some infectious type C retroviruses (MCF virus) suppresses lymphocyte activation in a murine model. Whether the described retroviral sequence in SLE patients and in other patients with autoimmune diseases and controls is indeed of some immunomodular activities remains to be proven. In this context, however, it should be mentioned that in other studies on plasma nucleic acids in SLE patients [37, 38] some characteristics similar to those plasma nucleic acids which were identified in our own studies with a high content of guanine-cytosine (GC) were obvious. This has been discussed by Sano and Morimoto [39] as being important for a possible change from the normal nonimmunogenic B-DNA to an immunogenic form. Characterization studies of clone base sequences from plasma DNA of SLE patients by Li and Steinman [40] suggest that the DNA in SLE patients contains a nonrandom selection of human genomic base sequences and is not of exogenous origin. However, the authors could not entirely exclude the possibility that a smaller fraction of this plasma DNA consists of a nonhuman genomic sequence.

Conclusions

From the discussed investigations it is clear that infectious and endogenous retroviruses can influence immunological mechanisms directly and indirectly, especially as shown in animal models. In the human situation, there is some indirect evidence that retroviruses might be involved in autoimmune and specifically rheumatic diseases. However, despite many years of investigations, a defined role of retroviruses in the pathogenesis of these disease entities still has to be established.

References

1. Brancato L, Itescu S, Skovron ML, Solomon G, Winchester R (1989) Aspects of the spectrum, prevalence and disease susceptibility determinants of Reiter's syndrome and related disorders associated with HIV infection. Rheumatol Int 9:137-141
2. Calabrese LH (1989) The rheumatic manifestations of infection with the human immunodeficiency virus. Semin Arthritis Rheum 18:225-239
3. Solinger AM, Adams LE, Friedman-Kien AE, Hess EV (1988) Acquired immune deficiency syndrome (AIDS) and autoïmmunity - mutally exclusive entities? J Clin Immunol 8:32-42
4. Mayer-Siuta R, Keil LB, DeBari VA (1988) Autoantibodies and circulating immune complexes in subjects infected with human immunodeficiency virus. Med Microbiol Immunol (Berl) 177:189-194
5. Bloom EJ, Abrams DI, Rodgers G (1986) Lupus anticoagulant in the acquired immunodeficiency syndrome. JAMA 256:491-493
6. DeClerk LS, Couttenye MM, DeBroe ME, Stevens WJ (1988) Acquired immunodeficiency syndrome mimicking Sjögren's syndrome and systemic lupus erythematosus. Arthritis Rheum 31:272-275
7. Berman A, Espinoza LR, Diaz JD et al. (1988) Rheumatic manifestations of human immunodeficiency virus infection. Am J Med 85:159-162
8. Saag MS, Bennett JC (1987) The infections etiology of chronic rheumatic diseases. Semin Arthritis Rheum 17:1-23
9. Krieg AM, Steinberg AD, Khan AS (1988) Increased expression of novel full-length endogenous MCF-related transcript in autoimmune mouse strains. Virology 162:274-276
10. Krieg AM, Khan AS, Steinberg AD (1989) Expression of an endogenous retroviral transcript is associated with murine lupus. Arthritis Rheum 32:322-329
11. Banks KL, Jutila MA, Jacobs CA, Michaels FH (1989) Augmentation of lymphocyte and macrophage proliferation by caprine arthritis-encephalitis

virus contributes to the development of progressive arthritis. Rheumatol Int 9:123-128

12. Kennedy-Stoskopf S (1989) Pathogenesis of lentivirus-induced arthritis. A review. Rheumatol Int 9:129-36
13. Weinhold KJ, Lyerly HK, Stanley SD, Austin AA, Matthews TJ, Bolognesi DP (1989) HIV-1 gp120-mediated immune suppression and lymphocyte destruction in the absence of viral infection. J Immunol 142:3091-3097
14. Mathes LE, Olsen G, Hebebrand LC, Hoover EA, Schaller JP (1978) Abrogation of lymphocyte blastogenesis by a feline leukemia virus protein. Nature 274:687-689
15. Hebebrand LC, Olson RG, Mathes LE, Nichols WS (1979) Inhibition of human lymphocyte mitogen and antigen response by a 15,000 dalton protein from feline leukemia virus. Cancer Res 39:443-447
16. Copelan EA, Rinehart JJ, Lewis M, Mathes L, Olson R, Sagone A (1983) The mechanism of retrovirus suppression of human T-cell proliferation in vitro. J Immunol 131:2017-2020
17. Chanh TC, Kennedy RC, Kanda P (1988) Synthetic peptides homologous to HIV transmembrane glycoprotein suppress normal human lymphocyte blastogenic response. Cell Immunol 111:77-86
18. Feldmann M, Ldonei M, Buchan G (1987) Interferons and autoimmunity. Interferon 9:75-90
19. Plotz PH (1983) Autoantibodies are anti-idiotype antibodies to antiviral antibodies. Lancet ii:824-826
20. Query CC, Keene JD (1987) A human autoimmune protein associated with U1RNA contains a region of homology that is cross-reactive with retroviral p30 gag antigen. Cell 51:211-220
21. Golding H, Robey FA, Gates III FT et al. (1988) Identification of homologous regions in human immuno-deficiency I gp41 and human MHC class II beta 1 domain. I. Monoclonal antibodies against the gp41-derived peptide and patients' sera react with native HLA class II antigens, suggesting a role for auto-immunity in the pathogenesis of acquired immune deficiency syndrome. J Exp Med 167:914-923
22. Rohdal E, Iversen O (1985) Antigens related to the major internal protein, p27, of a proriasis-associated retrovirus-like particle are expressed in patients with chronic arthritis. Ann Rheum Dis 44:761-765
23. Crawford TB, Adams DS, Cheevers WP, Cork LC (1980) Chronic arthritis in goats caused by a retrovirus. Science 207:997-999
24. Narayan O, Clements JE, Strandbertz JD et al. (1980) Biologic characterization of the virus causing leukoencephalitis and arthritis in goats. J Gen Virol 41:343-352.
25. Norval M, Hart H, Marmioin BP (1979) Viruses and lymphocytes in rheumatoid arthritis. I. Studies on cultured rheumatoid lymphocytes. Ann Rheum Dis 38:507:513
26. Gay S, Gay RE (1989) Cellular basis and oncogene expression of rheumatoid joint destruction. Rheumatol Int 9:105-113
27. Herrmann M, Leitmann W, Krapf FE, Kalden JR (1989) Molecular characteristics and in vitro effects of nucleic acids from plasma of patients with systemic lupus erythematosus. J Cell Biochem [Suppl] 105:147-157
28. Krapf F, Herrmann M, Leitmann W, Kalden JR (1989) Antibody binding of macromolecular DNA and RNA in the plasma of SLE patients. Clin Exp Immunol 75:336-342

29. Denman A (1986) Induction of interferon by lentiviruses. Immunol Today 7/8:201-202
30. Olsen RG, Tarr MJ, Mathes LE et al. (1987) Serological and virological evidence of human T-lymphotropic lymphotropic virus in SLE. Med Microbiol Immunol (Berl) 176/2:53-64
31. Talal N, Dauphinée MJ, Dang H, Alexander SS, Hart DJ, Garry RF (1990) Detection of serum antibodies to retroviral proteins in patients with primary Sjögren's syndrome (autoimmune exocrinopathy). Arthritis Rheum 33:774-781
32. Garry RF, Fermin CD, Hart DJ, Alexander SS, Donehower LA, Luo-Zhang H (1990) Detection of a human intracisternal A-type retroviral particle antigenically related to HIV. Science 250:1127-1129
33. Rucheton M, Graafland H, Fanton H, Ursule L, Ferrier P, Larsen CJ (1985) Presence of circulating antibodies against gag-gene MuLV proteins in patients with autoimmune connective tissue disorders. Virology 144:468-480
34. Krapf FE, Herrmann M, Leitman W, Kalden JR (1989) Are retroviruses involved in the pathogenesis of SLE? Evidence demonstrated by molecular analysis of nucleic acids from SLE patients' plasma. Rheumatol Int 9:115-121
35. Leitmann W, Herrmann M, Krapf E, Kalden JR (1990) Evidence for the involvement of retroviruses in the pathogenesis of systemic lupus erythematosus. In: Groopman JE, Chen ISY, Essex M, Weiss RA (eds) Human retroviruses. UCLA symposia on molecular and cellular biology, new series, vol 119. Wiley-Liss, New York, pp 361-372
36. Krieg AM, Gause WC, Gourley MF, Steinberg AD (1989) A role for endogenous retroviral sequences in the regulation of lymphocyte activation. J Immunol 143:2448-2451
37. McCoubrey-Hoyer A, Okarma TB, Holman HR (1984) Partial purification and characterization of plasma DNA its relation to disease activity in systemic lupus erythematosus. Am J Med 77:23-34
38. Sano H, Morimoto C (1982) DNA isolated from DNA/antiDNA antibody immune complexes in systemic lupus erythematosus is rich in guanine-cytosine content. J Immunol 128/3:1341-1345
39. Koga Y, Lindstrom E, Fenyo EM, Wigzell H, Mak TW (1988) High levels of heterodisperse RNAs accumulate in T-cells infected with human immunodeficiency virus and in normal thymocytes. Proc Natl Acad Sci USA 85:4521-4525
40. Li JZ, Steinman CR (1989) Plasma DNA in systemic lupus erythematosus. Characterization of clonal base sequences. Arthritis Rheum 32/6:726-733

Genetic Control of De Novo Conversion to Infectious Amyloids of Host Precursor Proteins: Kuru-CJD-Scrapie

D. Carleton Gajdusek

National Institute of Neurological Disorders and Stroke
National Institutes of Health, Bethesda, Maryland (USA)

Human spongiform encephalopathies, kuru and Creutzfeldt-Jakob disease (CJD), are caused by a virus that has entered the body years or decades earlier yet produces no disease until the final cascading fatal outcome [46,48]. The same is true of scrapie, the related disease of sheep and goats, which has served as a very good model for kuru [41,42,43,44,45,46,50,59]. It was already in the 1960s when we learned that the infectious agent of scrapie has a target size of its genetic information of less than 100 kDa, less than one-tenth that of any known virus, and that there was no translation and transcription involved in its replication since it did not contain a nucleic acid [4,5,6,49,60,77,78]. Knowing this, we yet continued to call these infectious agents viruses, as Robert Koch, Louis Pasteur, and Paul Ehrlich used the term "viruses" without implying anything about their biochemical composition. We keep to this original concept, as did the mathematicians when they called their nonmolecular computer parasites "viruses", meaning nothing more than invisible replicating parasites that required the energy and informational systems of the host for their replication. The infectious agents dealt with in this paper cause the human diseases kuru (a term for the disease from the Fore language in New Guinea meaning to shiver from fear or cold) [41,42,46,48], CJD and its variant Gerstmann-Sträussler-Scheinker syndrome (GSS) [81,97], and animal diseases such as scrapie and mink encephalopathy, wasting disease of mule deer and of elk, and bovine spongiform encephalopathy, all of which lead relentlessly to death [46].

Soon after the first transmission of scrapie was made [9,30,31,32], the unconventional properties and unusual stability of its agent were apparent. On my first visit to the Institute for Research on Animal Disease, Compton, England, I asked our British colleagues for their scrapie virus; they took it with a throat stick from a mucilage jar filled with mould and gave it to me on a piece of paper. In fact, some of the brain suspensions they stored were not kept in an ice-box but on an open shelf, even with an antiseptic added to prevent the growth of bacteria and yeasts.

Kuru as a Brain Amyloidosis

In the first histopathological study of kuru brains we found extracellular amyloid plaques which were metachromatic, PAS positive, and highly argentophilic; they were spherical and larger than neurons and green blue birefringent after staining with Congo red. We found them in 9 of the 12 brains first studied, all from preadolescent children [75]. The designation of such insoluble fibrillary extracellular precipitates as "amyloids" was a misnomer when it was first used. They were seen in many tissues, including the brain, and were thought to be composed of starch-like polysaccaride, hence the prefix "amyl". All amyloids are now known to be proteins, some glycosylated. In kuru we found these precipitates in more than 70% of our patients. This was the first time that such amyloid plaques were seen in the brains of children. However, nearly 25% of the brains, even after complete serial sectioning, did not show any amyloid plaques at all. Therefore we erroneously decided that the occurrence of these plaques was rather a side effect of the disease and not an indication of its primary pathogenesis.

Since most of the brains showed such plaques we began to call kuru a cerebral amyloidosis, and since such plaques were generally thought to be a senile change, we facetiously referred to kuru as a "galopping senescence of the juvenile". Clearly, the occurrence of amyloid plaques is the hallmark of the aging brain: 100% of the people who live to be 95 have such plaques in their brains; half of the people over 80 have them already, and some are found in 30% of brains of those over 70. Early in our study we prepared animal antisera to these plaques, which served to demonstrate them by immunofluorescence. We blundered in this early immunodemonstration in that we omitted to control the immune specificity with amyloid plaques from Alzheimer's disease, Down's syndrome, or the aging brain. Many studies were later done using scrapie virus and its amyloid plaques in mice because they were thought to be a good model for studying the pathogenesis of senile plaques in aging. The ultrastructure of the kuru and scrapie plaques showed amyloid fibrils identical to those of the amyloid in aging human brain [114]. It was not until the 1980s that we found that plaques from aging human brain were immunologically distinct from the scrapie and kuru plaques [3, 74], in spite of their morphological similarity and closely related histochemical properties [104]. These immunological studies demonstrated that kuru plaques and those in normal aging brain were not the same.

Amyloid plaques in all types of subacute spongiform virus encephalopathies showed the same immunoreactivity, but different from that of amyloid deposits in the aging brain, Alzheimer's disease, and Down's syndrome. Thus, the plaques found in brains from patients who died from Creutzfeldt-Jakob disease or its clinical variant, Gerstmann-Sträussler-Scheinker syndrome [81,104], and in animals suffering from the various spongiform encephalopathies (both those with naturally occurring disease and those infected experimentally) [7,18,37,40,57,58,63, 106] are composed of very similar molecules, yet are immunologically indistinguishable.

Composition of Plaques

The unconventional viruses causing mammalian spongiform encephalopathy have been better purified than most conventional viruses such as measles, herpesviruses, and retroviruses ever will be. Working on scrapie-infected brain tissues, Patricia Merz found in high titer density gradient banded preparations double stranded fibers seen by electron microscopy. These scrapie-associated fibrils (SAF) [84] are composed of insoluble, ß-pleated protein which we now know is a 27 kDa fragment of a normal host precursor protein of 35 kDa molecular weight, the scrapie precursor protein. It is this normal host precursor protein converted to an insoluble amyloid that is inextricably associated with scrapie infectivity. This subunit molecule of the SAFs and also of the amyloid fibrils in the scrapie amyloid plaques is a proteolytically cleaved 27 kDa fragment of the infectious amyloid form of the 35 kDa scrapie precursor protein. This 27 kDa fragment, if gently enough purified, is also infectious as is the full length infectious form of the scrapie precursor protein of 35 kDa (see Fig. 1) [98,99,100,101]

Conversion of the Normal Host Precursor Protein to an Infectious Molecule by Posttranslational Conformational Change

During the healthy life of the animal or human incubating the disease, the normal scrapie precursor protein is altered into an insoluble, protease-resistant form which is an amyloid by virtue of its post-translational conformational change to a ß-pleated sheet structure characterized by 7.6 Å and 10.6 Å spacing by X-ray crystallography. This form of the molecules can be isolated from lymph nodes, spleen and particularly brain, and the appearance of this protease resistant form is absolutely restricted to the subacute spongiform virus encephalopathies. Other degenerative diseases of the central nervous system and cerebrovascular accidents do not induce it. The normal and scrapie-modified form of the precursor can be separated by their different solubility properties: in different buffers, they can be separated on columns as two distinct bands [85]. Proteases remove the normal band. Amino acid sequencing [11] proved that the normal protein has the same amino acid sequence as the infectious protease resistant form. It is possible that the process of configurational change occurs extracellularly in the matrix, since the entire precursor protein is fully extruded and remains attached to the outer surface of the cellular membrane by a lipid anchor. All infectivity resides in this full-length configurationally changed precursor [28,98,99,100,101].

From this infectious 33-37 kDa host molecule a terminal piece can be split off by high activity proteinase K, leaving a still infectious 27 kDa polypeptide which enters gels and is easily polymerized [99,100,101]. This is the scrapie specific protein discovered by Prusiner and called his prion protein ($PrP_{27)}$ [94]. It may be eluted from gels and will polymerize in vitro into SAF-like rods, which Prusiner calls prion rods. At first, it appeared to be noninfectious. However, with gentler proteolysis it can be shown to retain the infectivity of the infectious form of the full-length precursor. We know now that both the full length ($PrP^{Sc}{}_{35}$) and cleaved (PrP_{27}) forms are infectious [28,101].

The normal, cellular form of the scrapie amyloid precursor protein has the form of a membrane anchored receptor molecule. It has a very rapid turn-over time of under 5 h in murine N2a neuroblastoma cells in culture, whereas the infectious scrapie amyloid form has a turnover time of over 15 h [12] because of its removal from proteolysis and accumulation. However, the important matter of its possible function has not yet been discovered. It has considerable homology with an acetylcholine receptor stimulating molecule found in chickens, but such function has not yet been documented for the normal scrapie precursor protein [62].

Resistance to Inactivation by Physical and Chemical Agents

Scrapie infectivity has low sensitivity to inactivation by UV at the optimal wave length of 250-256 nm for dimer formation [4,6,60,77,78]. The purified preparation can be treated with proteases, amylases, nucleases, or any other enzymes without changing the titer of infectivity [20]. Its extraordinary resistance to many chemical agents [34] but sensitivity to 5% hypochlorite (HClO) and to 1N NaOH have been reported [14,16,20].

What about temperature? Infectivity titers are slightly reduced by boiling the virus, and titers are reduced by 3 or 4 $\log_{10}$ by autoclaving under 15 atmospheres of pressure at 127°C, and after lyophilizing formaldehyde treated virus preparations and then heating them to 270°C for one hour in a dry oven all infectivity is preserved [21,22,107]. No infection nucleotides would withstand such temperatures. Two neurosurgeons and three neuropathological technicians [86,105,107] have come down with CJD. The current safety suggestion is to treat infected material with 1 N NaOH which is fully effective and cheap, and flush contaminated surfaces with 1 N NaOH or 5% hypochlorite [14,16,20].

Heat resistance, thus, provides strong evidence that there is no infective DNA or RNA in the infectious preparations of amyloid protein. Not only are the polymerized fibrils (SAF) infectious, but after destruction of the fibrils by sonication or detergents, the preparations remain infectious. Since infectious titers are many $\log_{10}$ lower there than the number of these 27 kDa amyloid monomer molecules in such inocula, it may be that the property of infectivity resides in a polymeric form containing many thousands of monomers.

The question remains how a normal endogenous protein can be converted into an infectious form. The many O and N linked glycosylation sites on the host precursor protein suggest that the glycosylation may have something to do with the infectivity. However, enzymatic deglycosylation of the infectious form of the protein fails to destroy the infectivity [20].

The full sequence of the scrapie precursor protein ($PrP^{C}{}_{35}$) has been deduced from a cDNA [79,90] and the gene for it has been localized on chromosome 20 in man and chromosome 2 in the mouse. We now have its total sequence in man, sheep, cow, mouse, rat and hamster. It is highly conserved [110], one of the most conserved proteins known. It is expressed in every cell of the body and exists in embryonic and adult life without extensive modulation. Studies of its mRNA by in situ hybridization have shown that it is not modulated during scrapie infection in the brain cells; it is not turned on. During infection it is post-translationally modified into a ß-pleated configuration of amyloid, which renders it protease resistant. This full length, 35 kDa infectious form is proteolytically cleaved at the one site still sensitive to proteases to become scrapie amyloid. The still infectious fragments readily polymerize to form the SAFs (see Fig. 1).

The full length 35 kDa infectious form displays no amino acid change as compared to the naturally occurring form [8,29]. The difference between the infectious form of the scrapie precursor protein and its noninfectious host precursor is not known and must eventually be resolved by X-ray crystallography or NMR techniques - but it is clearly a post-translational modification of the conformation of the normal host precursor proteins. It may even involve a covalent or coordinate covalent chemical bond formation, but not with any change in amino acid sequence [36,99,100,101]. Infectivity of the 35 kDa full length infectious form of the precursor proteins is preserved on boiling in SDS, which should destroy all H-bonded secondary, tertiary, or quaternary structure of a polypeptide chain. This may indicate a stronger chemical bonding than hydrogen bonds causing the conformational change in the precursor. The protease resistant, ß-pleated amyloid infectious form of the precursor (35 kDa) may be changed to a form no longer protease resistant by gentle treatment with lower concentration of SDS and yet preserve infectivity. Thus this has produced a protease sensitive infectious full length protein, the infectivity of which is quickly destroyed by proteases [36,101].

Nucleation of protein crystallization can be triggered by pulverized or ground minerals, as shown [83] for beef liver catalase, lysozyme, canavalin, and concanavalin B. Different minerals cause different patterns of crystal growth in the same protein, for example, bifurcating and linear forms, but possibly as many different patterns as in snowflakes or frost on a window. Any particles of the grown crystal are themselves able to act as pattern setting nucleants for further crystal formation. In scrapie infection it is likely that by a related process of pattern induction the infectious protein confers a change in structure on the normal precursor. The atomic, molecular, and crystallographic details of how such a process of autoinduced and patterned configurational change occurs remain to be elucidated. Whether chemical covalent or coordinate covalent bonding is catalyzed or only hydrogen bonds are altered to cause configurational change and polymerization will require elucidation at the quantum mechanical level. It is very unlikely that most neurons should undergo an alteration of biosynthesis of the precursor protein from altered gene transcription [26].

Incidence and Time Course of Human Transmissible Spongiform Encephalopathies: Kuru, CJD, and GSS

In the Fore tribe, kuru produced more than 90% of all deaths in adult women for a half century. During the first 20 of the 35 years of our surveillance it produced 50% of all deaths after infancy in the maximally affected villages; other causes of death in women between the ages of 12 and 30 were rare. In men the incidence was much lower, but for the whole population kuru presented a greater menace than the combined death rate from cancer, diabetes, cerebrovascular accidents, and coronary diseases with us. Killing more women than men, it changed the adult sex ration to 3:1 in much of the South fore region [41,42,46].

A direct chain of infection linked all cases of kuru; with the discontinuation of the opening of the Pandora's box crania containing the kuru-infected brains in the rite of ritual cannibalism, the disease has progressively disappeared. No person born since this cannibalistic contamination ceased over 30 years ago has developed kuru [46,76,95].

Kuru and CJD are molecularly identical in the amino acid sequence of their infectious amyloids, which are also immunologically identical [10,18]. In fact, they appear to be the same virus. CJD patients are often clinically distinguishable from Alzheimer's disease patients, but this is not always the case [16,82,97]. Alzheimer's disease is rarely fatal within 1 year of onset whereas CJD is usually fatal within 6 months of the onset and fewer than 5% of the cases survive for over 1 year. Few Alzheimer patients die even within 2 years of the onset. The mean age of CJD patients is in the late fifties, but among the over 3 000 CJD (and GSS) cases we have reviewed there was a patient as young as 11 years and others over 80 years of age.

When Kirschbaum reviewed CJD in 1968 [73] there had been only about 100 cases reported in the world since the first description in the 1920s. Our laboratory has now had almost 4 000 cases of CJD brougth to our attention. Interestingly, whenever we stimulated our colleagues in neurology to hunt for CJD patients in cities throughout the world, they usually quickly reached an incidence rate of one case per million population per annum. This has been the case in Japan and China, Chile and Argentina, and Australia and New Zealand. Intensive CJD surveillance in over a dozen countries has found an incidence everywhere of about one per million inhabitants per annum [19,82]. There is no other infectious disease with so uncannily a similar prevalence and incidence in all races and in all climates from the arctic to the tropics. One certainly wonders how an infectious disease should have the same incidence everywhere in the world. This strange finding first stimulated the conjecture over 30 years ago that the disease might result from a spontaneous somatic mutation (F.M. Burnet, personal communication) [46].

Truly sporadic cases of CJD account for over 85% of all cases. They have never had known contact with other CJD patients. Intensive effort to track down a possible source of infection has failed [19,82]. Strikingly, in Sephardic Jews of Libyan origin in Israel the disease was found at over 20 times the incidence in Ashkenazi Jews in Isreal. Foci of high incidence and local clustering of a few cases of CJD have been observed but only those of Sephardic Jews of Libyan origin in Isreal and in the Lucenec and Orava areas of Slovakia have been acceptable as a truly signigicant increased incidence [25,72,87,88]. It appears that familial occurrence could account for at least half of this high incidence, Sephardic Jews all belonging to an extended inbred family tree. We encounter here what we generally find in the familial type of CJD, namely, a really autosomal dominant pattern of expression. We have studied several dozen genealogies and could detect in all of these a true Mendelian dominant trait as impressive as that found in the genetic study of Huntington's disease. On the other hand, all of the affected individuals have the virus, as we proved by transmission to monkeys or other laboratory animals [81]. This, to my knowledge, is the only infectious disease of man in which expression of disease is proved to be controlled by a single autosomal dominant gene.

Attempts have been made to correlate the high incidence of the disease in these population groups with their habits of consuming sheep meat, especially brain and eye balls. On the other hand, the prevalence of CJD in professional groups such as shepherds, abattoir workers, and veterinarians does not exceed that among the normal population [19,27,41,42,82].

Extensive study of sporadic cases of CJD has failed to reveal a source of infection in the lifetime of the patients. They have never been known to have encountered another CJD patient. Thus, we have two unusual epidemiological findings for any infectious disease: first, the strangely uniform $1/10^6$ per year incidence and prevalence everywhere, and second, the lack of an infectious chain for the sporadic cases, except in the few iatrogenic cases where the virus has been inoculated parenterally and thus provided an established chain of infection. These two findings have led us to conjecture that the sporadic event may generate spontaneously their infectious amyloid by a rare stochastic event of their host precursor protein falling into the infectious ß-pleated configuration at an incidence rate of $1/10^6$ population per annum, the world-wide incidence of CJD. In the genetically determined cases of familial CJD and GSS the single amino acid substitution increases the likelihood of this rare event by about one million-fold [46].

The Problem of Host Range and Individual Susceptibility

Scrapie can be transmitted to primates by intracerebral inoculation. In Cynomolgus monkeys infected with the scrapie agent of sheep, the incubation period was over 6 years. In the rhesus monkey it was 13 years. The disease in monkeys resembles human CJD, and it can be transmitted back to goat and sheep. Transmission studies demonstrate that infection takes in any chimpanzee, squirrel, capuchin, or spider monkey into which it is inoculated, not only in certain selected individuals who have the right genetic constitution. Hence, in these species as in humans with kuru there is no genetic control of susceptibility in the inoculated host [76]. However, in contrast, familial CJD and GSS behave as a fully penetrant autosomal dominant trait.

Once we were aware that scrapie of sheep and goats can be transmitted to the mouse and the hamster we obtained a really workable system since, if one uses animals of the right genotype, one can get the disease in 3 or 4 months instead of 2 years. In those systems we can work fast. Some rare strains of the CJD agent infect cats, and some other strains infect guinea pigs and even mice. On the other hand, we have inoculated over 100 alley cats with dozens of CJD-infected human brains suspensions (which transmit to monkeys) and most of these animals never came down with the disease even after years of observation. Only about one in ten human strains are pathogenic for the cat. Yet, as far as we know, full single gene control of spongiform encephalopathy disease expression exists only in man.

The Transthyretin Amyloidoses of Familial Amyloidotic Polyneuropathy as a Paradigm for the Genetic Control of De Novo Genesis of CJD Infectious Amyloid by a Spontaneous Change in Configuration of the Host Precursor Protein

There is a large literature, unfortunately making no reference to any microbiology papers, dealing with the genetic control of amyloid genesis. Of most pertinence to our problem of the unconventional viruses is a group of diseases of the peripheral nervous system, called familial amyloidotic polyneuropathy (FAP). Patients are members of several dozen families from all around the world. The onset of the clinical disease occurs at different ages and leads to the destruction of peripheral nerves by progressive deposition of amyloid in the perineurium. These amyloid deposits, however, are not derived from the precursor of scrapie amyloid or the ß/A4 protein precursor of aging brain or Alzheimer's disease amyloid. Instead the FAP amyloids are derived from transthyretin, a prealbumin involved in the transport of thyroid hormone. The full sequence of this molecule is known and its encoding gene is located on chromosome 18 [102,108]. Transthyretin in its pure and crystalline form is a soluble prealbumin of 14 kDa molecular weight with 127 amino acids. It is amyloidogenic by structural chemical considerations, but amyloid formation does not apparently occur "on the hoof" - no sporadic cases with no mutation in the transthyretin gene are known. Members of the different affected families have a mutation resulting in a one amino acid substitution in the precursor that increases the statistical likelihood of the molecule falling into the amyloid form by a factor of about 10^4 to 10^6. There is not one specific mutation causing the disease in all families. Thus, in over two dozen investigated families, 17 different mutations have been detected. The familial amyloidotic polyneuropathy is caused by precipitation of amyloid formed from the transthyretin precursor, with any of a set of point mutations each causing a one amino acid replacement which increases the statistical likelihood of amyloid formation. This amyloid is not a replicating infectious molecule. Without one of these point mutations it is difficult to change the transthyretin polypeptide by concentration and nucleation into the amyloid configuration.

Protein	*Process*	*Synonymy*
scrapie precursor		$PrP^{C}{}_{33-35}$
↓	induction by autopatterning of configurational change	↓
infectious form of scrapie precursor		$PrP^{Sc}{}_{33-35}$
↓	proteolytic cleavage	↓
scrapie amyloid		PrP_{27}
↓	autonucleation of fibril polymerization	↓
scrapie-associated fibrils		prion rods

Fig. 1. De novo generation of infectious scrapie proteins from scrapie precursor

With these single amino acid substitutions amyloid formation is a much more likely stochastic event, even in vitro and extracellularly (see Fig. 2).

There are also several silent polymorphisms in the population with nonpathogenic single amino acid substitutions (Fig. 2). The codon 30 point mutation with proline replaced by methionine has been expressed in transgenic mice which develop deposits of amyloid containing the methionine mutation similar to depositions in familial amyloidotic polyneuropathy, but also in the intestine and other tissues, and pass this trait to their offspring [109,113].

Parallels to familial CJD and its familial GSS variant are obvious (see Fig. 3). Most GSS families display an amino acid replacement of proline by leucine at codon 102 [56,65,69]. In a family with atypical GSS there is instead a replacement of alanine by valine at codon 117 [38,68]. Other GSS families have none of these mutations [67]. The more common type of familial CJD has a codon 200 mutation which replaces glutamic acid with lysine [51,52,53,54,56]. This has now been found in 14 families. However, in a large Finnish kindred with CJD [55] there is a replacement of aspartic acid by asparagine in codon 178, and there are Dutch, French, Hungarian and American CJD families also with codon 178 point mutation [89]. British colleagues investigating a pedigree with CJD patients in America and England have found a 48 amino acid insert, an octapeptide six-fold repeat, between codons 52 and 91 [91,92,93]. Thus we know five different mutations which are responsible for the alteration of the normal precursor protein into the infectious amyloid form. At codon 129 we have a nonpathogenic substitution point mutation which is a silent polymorphism in the general population [51].

At present the best explanation for the regular incidence of sporadic nonfamilial CJD around the world is the de novo creation of the CJD amyloid infectious agent by a rare, spontaneous event occurring at a frequency of one per million population per annum, the world-wide incidence of CJD [46]. If one of the point mutations of familial CJD is present this configurational change occurs with about a millionfold higher likelihood. One corollary of this paradigm that has already been proved is that, although infectious, the replication of the amyloid caused by inoculations of a different individual does not "breed through". The point mutation is not copied in the amyloid formed in the new host, although it, in turn, is also infectious. The new CJD amyloid has the amino acid sequence of the newly inoculated individual. CJD from patients with the 102, 117, 178, and 200 codon mutations have all been transmitted to monkeys or chimpanzees which do not carry these point mutations, nor do the infectious proteins made in these experimentally infected hosts contain those point mutations.

Therefore these rare, so-called unconventional slow virus diseases present several heresies to those with implicit faith in the central dogma of modern molecular biology. First, they spontaneously create themselves; second, they contain no nucleic acid and thus new copies of themselves are made without any transcription or translation; third, in their replication in a new individual they do not copy any mutation in the nucleating and pattern setting agent. One implication of this is that if one inoculates goats with human brain material infected with CJD or kuru, the goat will develop the disease, but the sequence of its "virus" is not that of man but of the goat. If the agent is transmitted from goat to sheep, the sheep sequence will be obtained. The goat is susceptible to kuru, to CJD and also to mink encephalopathy (a disease that occurred in the Soviet

Union, Finland, Sweden, Canada, the United States, and Japan from feeding carcasses of scrapie-infected sheep to mink). If a mink is inoculated with sheep scrapie, the resulting mink virus has the amino acid sequence of the scrapie precursor protein of mink, not that of the sheep. The mink virus, however, is no longer infectious in the mouse, even if it originally came from a mouse passage of sheep scrapie. But, after passage through a goat, it will infect the mouse. These changes of host range appear to be mutations, and 5 years ago everyone would have considered them mutations, but they are, in fact, induced post-translational configurational changes in the precursor with the amino acid sequence of the normal precursor protein maintained.

Scrapie, Bovine Spongiform Encephalopathy and Some Putative Implications

Scrapie has been endemic probably for centuries in England, France, Hungary, and Czechoslovakia and is surely more widespread than generally assumed. In France, officials had told us that, in spite of the existence of many French laboratories working on these viruses, the natural disease (in French "*tremblante*") had disappeared. Scrapie was first reported to be transmissible in France in 1899 by Besnoit [9] and later proved by Cuillé and Chelle [30,31,32]. In the 1970s and early 1980s, there were no reports of scrapie in France. In 1980 we searched for scrapie in France and found hundreds of cases throughout the country entering the food chain as occurs elsewhere [27]. We were not able to relate the appearance of CJD to the consumption of lamb or mutton or to workers with sheep on farms or abattoirs. Three decades of intensive CJD epidemiology in France have failed to establish any association with scrapie [19].

Fig. 2. Sixteen different amino acid substitutions caused by point mutations in the gene specifying the transthyretin prealbumin amyloid precursor molecule in over 20 families of various ethnic origin are shown. Four of these families are normal without FAP, and the mutation is a silent nonpathogenic polymorphism in these. On codon 49 are two different amino acid substitutions, in a Jewish and Italian family, respectively. Codon 90 (his→asn) mutation has apparently caused FAP in the Italian:Sicilian family, but not in the German and Portuguese families

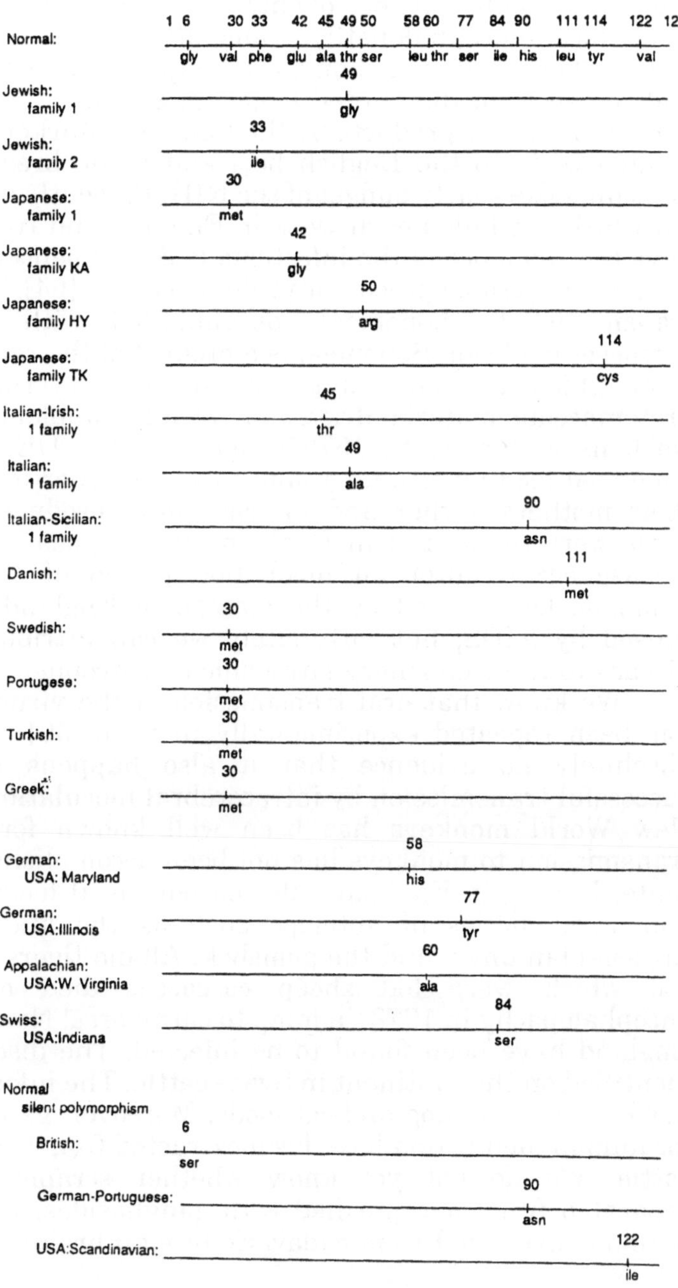
TRANSTHYRETIN AMYLOIDOSES OF FAMILIAL AMYLOIDOTIC POLYNEUROPATHY (FAP)
Mutations Increasing Likelihood of Host Precursor
Falling Into Amyloid Configuration
1 6 30 33 42 45 49 50 58 60 77 84 90 111 114 122 127
Normal:
gly val phe glu ala thr ser leu thr ser ile his leu tyr val
Jewish: family 1
49 gly
Jewish: family 2
33 ile
Japanese: family 1
30 met
Japanese: family KA
42 gly
Japanese: family HY
50 arg
Japanese: family TK
114 cys
Italian-Irish: 1 family
45 thr
Italian: 1 family
49 ala
Italian-Sicilian: 1 family
90 asn
Danish:
111 met
Swedish:
30 met
Portugese:
30 met
Turkish:
30 met
Greek:
30 met
German: USA: Maryland
58 his
German: USA:Illinois
77 tyr
Appalachian: USA:W. Virginia
60 ala
Swiss: USA:Indiana
84 ser
Normal silent polymorphism
British:
6 ser
German-Portuguese:
90 asn
USA:Scandinavian:
122 ile

Cows have been living in close contact with sheep for several thousands of years in many places of the world, but it has been the questionable privilege of only the present decade to recognize spongiform encephalopathy transmitted from sheep to cattle [2,112]. The world-wide ban on the importation of breeding stock and semen from the United Kingdom and since November 1989 of beef and dairy products by the Common Market countries has created costs to the English beef and cattle breeding industry exceeding the yearly budget of the NIH. Presently about 500 cows with BSE are killed each week in England and Ireland. We now have the sequence of the infectious BSE polypeptide; it is that of the normal scrapie precursor protein of cattle [64]. The agent was transmitted by bonemeal contaminated with scrapie from carcasses of sheep. Bonemeal is a product of the rendering plants from which the lard is taken off and the remaining high proteinaceous material dried and given as supplementary feed to chickens and pigs. In 1978 farmers in the UK began to use bonemeal feeding also for young calves in order to take them off their mothers earlier and increase milk production. The first cases were recognized in England 1984. Appearance of clinical disease occurs after an incubation period of 4-8 years. The economic loss caused by the disease in England exceeds that caused by AIDS, but fortunately we can attribute no human disease to this veterinary iatrogenic catastrophe.

We know that oral transmission of the virus occurs - this has been repeated experimentally in mice [39] - but we have absolutely no evidence that it also happens in man. The successful transmission by intracerebral inoculation of scrapie to New World monkeys has been well known for years. Such transmission to monkeys has not been accomplished by the oral route. In zoos in England, infected bonemeal feeding has killed four rare species of antelope such as the kudu, eland, the Abyssianian onyx, and the gemsbok. Albino tigers in the Bristol zoo which were fed sheep carcasses died of spongiform encephalopathy in 1978 - a jump to carnivores! Now many cats in England have been found to be infected. The disease has been identified on the continent in Swiss cattle. The infected bonemeal has been used in dog and cat foods. Moreover, gangliosides used for human medication have been extracted from brains of British cattle. We do not yet know whether scrapie infectivity is separated from or copurified with gangliosides. Human growth hormone extracted from cadaveric human brains, some of them

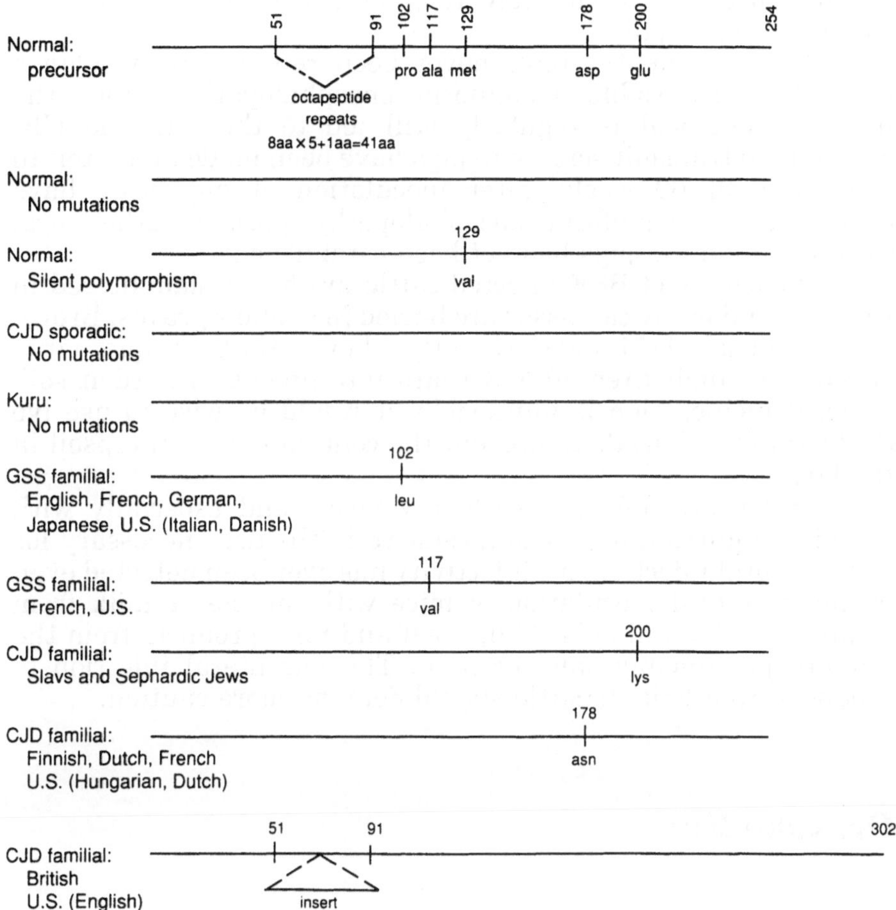

Fig. 3. Five different amino acid substitutions caused by point mutations in the gene specifying the host precursor molecule CJD amyloid. Four are found in families of diverse ethnic origin with familial CJD and its GSS variant. The fifth is the codon 129 substitution of valene for methionine, which is a silent polymorphism found in about 20% of the normal population. A sixth mutation in three families with CJD is an octapeptide repeat, with six copies (8aa x 6 = 48aa) inserted into a region where there are already five copies of the same repeat, bringing the total to eleven copies. There are also families of familial CJD and GSS without any point mutations

infected with CJD, was so thoroughly separated from the virus by its purification that only minimal infectivity remained. Thus only 14 cases of the disease have occurred, instead of a much higher number in the approximately 100 000 human growth hormone treated patients [13].

Pigs and poultry have never been reported to develop a natural transmissible spongiform encephalopathy. Since the infected bonemeal is regularly still fed to them in the UK, attempts to transmit scrapie to pigs have been made and after an incubation of 69 weeks after inoculation of pigs they have developed a spongiform encephalopathy [33]. Whether oral infection occurs in pigs has not been established.

Thousands of BSE infected cattle are being slaughtered in England and many carcasses are buried in shallow graves. Brown and Gajdusek [15] have recently shown that scrapie virus survives in high titer after 3 years internment in garden soil. Thus, if incineration is impossible, it would be wise to use lye (NaOH) in the burials to prevent the contamination of topsoil in the UK.

Oral route infection with any virus, and especially with scrapie, requires huge inocula relative to the dose necessary for intracerebral infection. No infectivity has ever been detected even by intracerebral inoculation of mice with muscle or milk from scrapie-infected animals. Thus, beef and milk products from the UK are presumably safe for man. The parenteral injection of products from British cattle should demand more caution.

Oravske Kuru

We now have a new focus of CJD in high incidence in man reported in Slovakia by Mitrová [87]. She has been following the incidence of CJD throughout Slovakia for more than two decades. In 1980 she identified an unusually high incidence of CJD in the rural Lucenec area of south central Slovakia with many cases also across the border in Hungary [87]. During the past decade cases have been found in increasing frequency from the most sparsely populated area of Slovakia, Orava, to the west of the High Tatra mountains on the Polish border. Here an epidemic of

CJD has developed during the 1980s with some 30 cases occurring in patients born and reared in a dozen small rural villages with a total population of under 15 000. This yields an incidence over 1 000 per million population per year in contrast to the worldwide incidence of one per million per year. The most intensely affected villages of Zuberec and Habovka with a total population of under 2 000 have had over 20 cases of CJD in the past 3 years. The incidence in the villages has thus been over 3 000 times higher than in the rest of the world in such cities as Paris, Berlin, Boston, New York, Sydney, Santiago, Beijing, or any other large cities, and it is still 100 times higher than that among the Sephardic Jews in Isreal. Members of the same family who were 20 to 30 years different in age became sick at nearly the same time. This suggested a common source infection rather than genetically determined etiology, as also did the new "epidemic" of appearance of CJD in the 1980s. For these reasons at first we believed that this outbreak could not be explained genetically.

However, we have now sequenced DNA from 9 of the Orava and 6 of the Lucenec CJD brains, and all have shown the substitution of lysine for glutamic acid in codon 200. Four of the 11 healthy adult first-order relatives studied have the same mutation. CJD had not been known in Orava before the 1970s [87]. The epidemic started with a few cases in the late 1970s and has developed into an escalating epidemic in the late 1980s. We have found some family members with the mutation although they are healthy and over 70 years of age. We are now looking for the cofactor that turns on the expression of the mutation, or a factor which in the past inhibited the post-translational configuration change of the precursor to amyloid. Thus, the new question is not what has caused the Orava outbreak - it is the codon 200 glutamic acid to lysine point mutation - but rather what has prevented its expression in previous generations so that it has accumulated as a frequent silent nonpathogenic polymorphism, only expressing itself as a pathogenic mutation in these people in the past 15 years?

The CJD Genetic Marker for the Wandering Jews of the Diaspore

On discovering the codon 200 glutamine to lysine point mutation responsible for the high incidence foci of CJD in both the Lucenec and Orava regions of Slovakia and widely disseminated in slavic peoples of Eastern Europe, we screened a large number of sporadic and familial CJD brain specimens from our frozen brain archive accumulated over the past 30 years [53]. This led us to discover the mutation in Greek CJD patients who were Sephardic Jews and quickly we found the mutation in Sephardic Jews who had come for diagnosis of CJD in France from Tunisia and in Sephardic Jews with CJD in Israel, both Libyan-born and Israel-born. Ashkenazic Jewish CJD patients did not have the codon 200 glutamine to lysine point mutation [55].

We are thus now investigating other Circum-Mediterranean Sephardic Jews with CJD, paying particular attention to the Iberian Peninsula, particularly Spain where in 1492 the Catholic monarchs Ferdinand and Isabella forced the quick conversion of large numbers of Sephardic Jews to Catholicism. Many of the remainder fled and gave rise to the large Sephardic Jewish group in Greece, where we have found the mutation.

One of the largest familial CJD pedigrees is that published by Haltia et al. [61] in Finland, which we have now investigated and found therein none of the point mutations previously known in familial CJD or GSS, but instead a codon 178 replacement of aspartic acid by asparagine. We have now found this codon 178 mutation in Dutch, French, Hungarian, and American cases of familial CJD [89].

Thus, our use of the paradigm of multiple point mutations causing the enormously increased likelihood of a post-translational conversion of the host precursor molecule to an amyloid configuration and deposition of insoluble amyloid in various tissues that we found in the FAP literature has proved amazingly predictive in our unravelling of the familial CJD and GSS pathogenesis.

Dependence of Host Range, Incubation Period, Plaque Production, and Lesion Distribution on Mutation in the Precursor Protein Gene

Primary transmission of any of these infections to a new host usually only affects a minority of the infected animals and only after a prolonged incubation period. This species barrier may disappear on one further passage in the new host, and the incubation period is usually shortened. The incubation period may vary greatly in different breeds in mice, hamsters, or sheep, as does also the appearance of amyloid plaques. These breed differences are associated with different sequences in the scrapie precursor protein.

In mice point mutations in the scrapie precursor protein determine the incubation period [23,24,111]. In different breeds of hamsters, incubation periods are similarly determined. Chinese and Syrian hamsters which develop amyloid plaques have mutations in codons 102 and 107, respectively, which are not found in Armenian hamsters which do not develop plaques [80]. Transgenic mice expressing hamster scrapie precursor gene develop hamster specific amyloid plaques and distribution of these plaques on hamster scrapie inoculation. They express mostly hamster scrapie infectious amyloid protein and much less infectious mouse amyloid protein. When inoculated with mouse scrapie they express only mouse scrapie infectious amyloid and no plaques [96,103]. In different breeds of sheep the inoculation period depends similarly on the amino acid sequence in the precursor protein [71].

Expression of the CJD Precursor Protein with These Point Mutations in Transgenic Mice and Baculovirus Systems and Synthetic Polypeptides with These Mutations

Now, as with FAP, the GSS point mutation planted in transgenic mice results in the production of pathology similar to that in the human disease. Hsiao's [66,70] recent implantation of the 102 codon proline to leucine mutated gene in transgenic mice with resulting spongiform disease in the mice is extremely encouraging; if the configurationally altered human precursor in these mice is an infectious amyloid our understanding will be enormously advanced.

We have now expressed these point mutations in the human scrapie precursor molecules in Baculovirus-cloned human precursor molecules and also synthesized large portions of the human homologue of the scrapie precursor with the appropriate point mutations. If these synthetic polypeptides or the Baculovirus-cloned proteins do serve as nucleants to produce the infectious transmission, the quick unravelling of the molecular event on an atomic level that produce this infectious change may be anticipated.

References

1. Adams DH (1972) The scrapie agent: a small desoxiribonucleid acid mediated virus. Biochem J 127:82-83
2. Aldhous P (1990) BSE causing public alarm. Nature 343:196
3. Allsop D, Ikeda S, Bruce M, Glenner GG (1988). Cerebrovascular amyloid in scrapie-affected sheep reacts with antibodies to prion protein. Neurosci Lett 92:234-239
4. Alper T, Cramp WA, Haig DA, Clark MC (1967) Does the agent of scrapie replicate without a nucleic acid? Nature 214:764-766
5. Alper T, Haig DA, Clarke MC (1966) The exceptionally small size of the scrapie agent. Biochem Biophys Res Commun 22:278-284
6. Alper T, Haig DA, Clarke MC (1978) The scrapie agent: Evidence against its dependence of replication of intrinsic nucleic acid. J Gen Virol 41:503-506
7. Bahmanyar S, Williams ES, Johnson FB, Young S, Gajdusek DC (1985) Amyloid plaques in spongiform encephalopathy of mule deer. J Comp Pathol 95:1-5
8. Basler K, Oesch B, Scott M, Westaway D, Wälchli M, Groth DF, et al. (1986) Scrapie and cellular PrP isoforms are encoded by the same gene. Cell 46:417-428
9. Besnoit C (1899) La tremblante ou nevrite peripherique enzootique du mouton. Rev Vet Toulouse 24:265-277, 333-343
10. Bode L, Pocchiari M, Gelderblom H, Diringer H (1985) Characterization of antisera against scrapie-associated fibrils (SAF) from affected hamster and cross-reactivity with SAF from scrapie affected mice and from patients with Creutzfeldt-Jakob disease. J Gen Virol 66:2471-2478

11. Bolton DC, Bendheim PE, Marmorstein AD, Potempska A (1987) Isolation and structural studies of the intact scrapie agent protein. Arch Biochem Biophys 258:579-590
12. Borchelt DR, Scott M, Taraboulos A, Stahl N and Prusiner SB (1990) Scrapie and cellular prion proteins differ in their kinetics of synthesis and topology in cultured cells. J Cell Biol 110:743-752
13. Brown P (1988) The decline and fall of Creutzfeldt-Jakob disease associated with human growth hormone therapy. Neurology 38:1135-1187
14. Brown P (1990) Guidelines for high risk autopsy cases: special precautions for Creutzfeldt-Jakob disease. In: autopsy performance and reporting. College of American Pathologists. Northfield, Illionois, pp 68-74
15. Brown P, Gajdusek DC (1991) Scrapie virus survives a three-year internment. Lancet 337:269-270
16. Brown P, Gibbs CJ Jr, Amyx HL, Kingsbury DT, Rohwer RG, Sulima MP, Gajdusek DC (1982a) Chemical disinfection of Creutzfeldt-Jakob disease virus. N Engl J Med 306:1279-1282.
17. Brown P, Salazar AM, Gibbs CJ Jr, Gajdusek DC (1982b) Alzheimer's disease and transmissible virus dementia (Creutzfeldt-Jakob disease). Ann NY Acad Sci 396:131-143
18. Brown P, Coker-Vann M, Pomeroy K, Franko M, Asher DM, Gibbs CJ Jr, Gajdusek DC (1986) Diagnosis of Creutzfeldt-Jakob disease by Western blot identification of marker protein in human brain tissue. N Engl J Med 314:547-551
19. Brown P, Cathalá F, Raubertas RF, Gajdusek DC and Castaigne P (1987) The epidemiology of Creutzfeldt-Jakob disease: conclusion of a 15 year investigation in France and review of the world literature. Neurology 37:895-904
20. Brown P, Liberski PP, Wolff A, Gajdusek DC (1990a) Conservation of infectivity in purified fibrillary extract of scrapie infected polyacrylamide gel electrophoresis. PNAS 87:7240-7244.
21. Brown P, Wolff A, Gajdusek DC (1990b) A simple and effective method for inactivating virus infectivity in formalin-fixed samples from patients with Creutzfeldt-Jakob disease. Neurology 40:887-890
22. Brown P, Wolff A, Liberski PP, Gajdusek DC (1990c) Resistance of scrapie infectivity to steam autoclaving after formaldehyde fixation, and limited survival after ashing at 306°C: practical and theoretical implications. J Inf Dis 161:467-472
23. Carlson GA, Kingsbury DT, Goodman PA, Coleman S, Marshall ST, De Armond S, et al. (1986) Linkage of prion protein and scrapie incubation time genes. Cell 46:503-511
24. Carlson GA, Goodman PA, Lovett M, Taylor BA, Marshall ST, De Armond S, et al. (1988) Genetics and polymorphism of the mouse prion gene complex: The control of scrapie incubation time. Mol Cell Biol 8:5528-5540
25. Cathalá F, Brown P, LeCanuet P, Gajdusek DC (1985) High incidence of Creutzfeldt-Jakob disease in North African immigrants to France. Neurology 35:894-895
26. Caughey B, Race RE, Chesebro B (1988) Detection of prion protein mRNA in normal and scrapie-infected tissues and cell lines. J Gen Virol 69:711-716
27. Chatelain J, Cathalá F, Brown P, Raharison S, Court L, Gajdusek DC (1981) Epidemiologic comparisons between Creutzfeldt-Jakob disease and scrapie in France during the 12-year period 1968-1979. J Neurol Sci 51:329-337

28. Ceroni M, Piccardo P, Safar J, Gajdusek DC, Gibbs CJ Jr (1990) Scrapie infectivity and prion protein are distributed in the same pH range in agarose isoelectric focusing. Neurology 40:508-513
29. Chesebro B, Race R, Wehrly K, Nishio J, Bloom M, Lechner D, et al. (1985) Identification of scrapie prion protein-specific mRNA in scrapie-infected and uninfected brain. Nature 315:331-333
30. Cuillé J, Chelle PL (1936) Pathologie animal - la maladie dite tremblante du mouton est-elle inoculable? CR Acad Sci 203:1552-1554
31. Cuillé J, Chelle PL (1939a) Transmission experimentale de la tremblante a la chévre. CR Acad Sci 208:1058-1060
32. Chuillé J, Chelle PL (1939b) Investigations of scrapie in sheep. Vet Med 34:417-418
33. Dawson M, Wells GAH, Parker BNJ, Scott AC (1990) Primary parenteral transmission of bovine spongiform encephalopathy to the pig. Vet Rec 127:338
34. Dickinson AG, Taylor DM (1978) Resistance of scrapie agent to decontamination. New Eng J Med 229:1413
35. Diringer H (1990) Laboratory markers for detection of scrapie. J Am Vet Med Assoc 196:1683
36. Diringer H, Blode H, Oberdick U (1991) Virus-induced amyloidosis in scrapie involves a change in covalent linkages in the preamyloid. Arch Virol (Vienna), in press
37. Doh-ura K, Tateishi J, Kitamoto T, Sasaki H, Sakaki Y (1989a) Creutzfeldt-Jakob disease patients with congophilic kuru plaques have the missense variant prion protein common to Gerstmann-Sträussler syndrome. Ann Neurol 27:121-126
38. Doh-ura K, Tateishi J, Sasaki H, Kitamoto T, Sakaki Y (1989b) Protein change at position 102 of prion protein gene is the most common but not the sole mutation related to Gerstmann-Sträussler syndrome. Biochem Biophys Res Communicat 163:974-979
39. Fraser H, McConnell I, Wells GAH, Dawson M (1988) Transmission of bovine spongiform encephalopathy to mice. Vet Rec 123:472
40. Fukatsu R, Gibbs CJ Jr, Gajdusek DC (1984) Cerebral amyloid plaques in experimental murine scrapie. In: Tateishi J (ed) Proceedings of the workshop on slow transmissible diseases. Research Committee on Slow Virus Infections, Japanese Ministry of Health, 31. August 1984, Tokyo, 71-84
41. Gajdusek DC (1976) Unconventional viruses and the origin and disappearance of kuru. In: Les Prix Nobel en 1976, Nobel Foundation, Norstedt, Stockholm, pp 167-216
42. Gajdusek DC (1977) Unconventional viruses and the origin and disappearance of kuru. Science 197:943-960
43. Gajdusek DC (1988a) Transmissible and non-transmissible amyloidoses: autocatalytic post-translational conversion of host precursor proteins to ß-pleated configurations. J Neuroimmunol 20:95-110
44. Gajdusek DC (1988b) Etiology versus pathogenesis: the causes of posttranslational modifications of host specified brain proteins to amyloid configuration. In: Sinet PM, Lamour Y, Christen Y (eds) Genetics and Alzheimer's disease. Proceedings of a meeting held by the foundation IPSEN pour la Recherche Thérapeutique, Paris, March 25. Springer, Berlin, pp 174-176

45. Gajdusek DC (1989) Fantasy of a "virus" from the inorganic world: pathogenesis of cerebral amyloidoses by polymer nucleating agents and/or"viruses". In: Neth R, Gallo RC, Greaves MF, et al. (eds) Modern trends in human leukemia VIII. Springer, Berlin, Heidelberg, New York, pp 481-499
46. Gajdusek DC (1990) Subacute spongiform encephalopathies: transmissible cerebral amyloidoses caused by unconventional viruses. Chapter 20 in: Fields BN, Knipe DM, Chanock RM, Hirsch MS, Melnick JL, Monath TP, Roizman B (eds) Virology, 2nd ed, Raven, New York, pp 2289-2324
47. Gajdusek DC, Gibbs CJ Jr (1990) Brain amyloidoses: precursor proteins and the amyloids of transmissible and nontransmissible dementias: Scrapie-kuru-CJD viruses as infectious polypeptides or amyloid-enhancing factors.In: Goldstein AL (ed) Biomedical advances in aging, Plenum, New York, pp 3-24
48. Gajdusek DC, Zigas V (1957) Degenerative disease of the central nervous system in New Guinea. The endemic occurrence of "kuru" in the native population. New England Med 257:974-978
49. Gibbs CJ Jr, Gajdusek DC, Latarjet R (1978) Unusual resistance to ionizing radiation of the viruses of kuru, Creutzfeldt-Jakob disease, and scrapie. Proc Natl Acad Sci USA 75:6268-6270
50. Gibbs CJ Jr, Safar J, Ceroni M, di Martino A, Clark WW, Hourrigan JL (1990) Experimental transmission of scrapie to cattle. Lancet 335:1275
51. Goldfarb LG, Brown P, Goldgaber D, Asher D, Rubenstein R, Brown WT, et al. (1990a) Creutzfeldt-Jakob disease and kuru patients lack a mutation consistently found in Gerstmann-Sträussler-Scheinker syndrome. Exp Neurol 108:247-250
52. Goldfarb LG, Brown P, Goldgaber D, Farrato RM, Yanagihara R, Asher DM, Gajdusek DC (1990b) Identical mutation in unrelated patients with Creutzfeldt-Jakob disease. Lancet 336:174-175
53. Goldfarb LG, Korczyn AO, Brown P, Chapman J, Gajdusek DC (1990c) Mutation in codon 200 of scrapie amyloid precursor gene linked to CJD in Sephardic Jews. Lancet 336:637
54. Goldfarb LG, Mitrová E, Brown P, Toh BH, Gajdusek DC (1990d) Mutation in codon 200 of scrapie amyloid protein gene in two clusters of Creutzfeldt-Jakob disease in Slovakia. Lancet 336:514-515
55. Goldfarb LG, Haltia M, Brown P, Nieto A, Kovena J, McCombie WR, et al. (1991) New mutation in the scrapie amyloid precursor gene (at codon 178) in Finnish Creutzfeldt-Jakob disease kindred. Lancet 337:1
56. Goldgaber D, Goldfarb LG, Brown P, Asher DM, Brown WT, Lin S, et al. (1989) Mutations in familial Creutzfeldt-Jakob disease and Gerstmann-Sträussler syndrome. Exp Neurol 106:204-206
57. Guiroy DC, Williams ES, Yanagihara R, Gajdusek DC (1990a) Immunolocalization of scrapie amyloid in chronic wasting disease in Rocky Mountain elk and hybrids of captive mule deer and white-tailed deer. Neurosci Lett, in press
58. Guiroy DC, Williams ES, Yanagihara R, Gajdusek DC (1990b) Topographic distribution of scrapie amyloid-immunoreactive plaques in chronic wasting disease in captive mule deer (Odocoileus hemionus hemionus). Acta Neuropathol, in press
59. Hadlow WJ (1959) Scrapie and kuru. Lancet 2:289-290

60. Haig DC, Clark MC, Blum E, Alper T (1969) Further studies on the inactivation of the scrapie agent by ultraviolet light. J Gen Virol 5:455-457
61. Haltia M, Kovanen J, von Crevel H, Bots GTHAM, Stefanko S (1979) Familial Creutzfeldt-Jakob disease. J Neurol Sci 42:381-389
62. Harris DA, Falls DL, Johnson FA, Fischbach GD (1991) A chick prionlike protein co-purifies with an activity that promotes the synthesis of acetylcholine receptors in cultured myotubes. Science, in press
63. Hope J, Multhaup G, Reekie LJ, Kimberlin RH, Beyreuther K (1988a) Molecular pathology of scrapie-associated fibril protein (PrP) in mouse brain affected by the ME7 strain of scrapie. Eur J Biochem 172:271-277
64. Hope J, Reekie LJ, Hunter N, Multhaup G, Beyreuther K, White H, et al. (1988b) Fibrils from brains of cows with new cattle disease contain scrapie-associated protein. Nature 336:390-392
65. Hsiao K, Baker HF, Crow TJ, Poulter M, Owen E, Terwilliger JD, et al. (1989) Linkage of prion protein missense variant to Gerstmann-Sträussler syndrome. Nature 338:342-345
66. Hsiao KK, Prusiner SB (1990) Inherited human prion diseases. Neurology 40:1820-1827
67. Hsiao K, Cass C, Conneally PM (1990a) Atypical Gerstmann-Sträussler-Scheinker syndrome with neurofibrillary tangles: No mutation in the prion protein open-reading-frame in a portion of the Indiana kindred. Neuro Biol Aging 11:3, 302
68. Hsiao K, Cass C, Schellenberg G, Bird T, Devine, Gage E, et al. (1990b) A prion protein variant in a family with a telencephalic form of Gerstmann-Sträussler-Scheinker syndrome. Neurology, in press
69. Hsiao KK, Doh-ura K, Kitamoto T, Tateishi J, Prusiner SB (1990c) A prion protein amino acid substitution in ataxic Gerstmann-Sträussler syndrome. Ann Neurol 26:137
70. Hsiao K, Scott M, Foster D, Groth DF, DeArmond SJ, Prusiner SB (1990d) Spontaneous neurodegeneration in transgenic mice with mutant prion protein. Science 250:1587-1590
71. Hunter N, Foster JD, Dickinson AG, Hope J (1989) Linkage of the gene for the scrapie-associated fibril protein (PrP) to the *Sip* gene in Cheviot sheep. Vet Rec 124:364-366
72. Kahana E, Alter M, Braham J, Sofer D (1974) Creutzfeldt-Jakob disease: focus among Libyan Jews in Israel. Science 183:90-91
73. Kirschbaum WR (1968) Creutzfeldt-Jakob disease. American Elsevier, New York
74. Kitamoto T, Tateishi J, Tashina T, Takeshita I, Barry RA, DeArmond SJ, Prusiner SB (1986) Amyloid plaques in Creutzfeldt-Jakob disease strain with prion protein antibodies. Ann Neurol 20:204-208
75. Klatzo I, Gajdusek DC, Zigas V (1959) Pathology of kuru. Lab Invest 8:799-847
76. Klitzman RL, Alpers MP, Gajdusek DC (1984) The natural incubation period of kuru and the episodes of transmission in three clusters of patients. Neuroepidemiology 3:3-20
77. Latarjet R (1979) Inactivation of the agents of scrapie, Creutzfeldt-Jakob and Kuru by radiations. Slow transmissible diseases of the nervous system. Academic, New York

78. Latarjet R, Muel B, Haig DA, Clarke MD, Alper T (1970) Inactivation of the scrapie agent by near-monochromatic ultraviolet light. Nature 227:1341-1343
79. Locht C, Chesebro B, Race R, Keith JM (1986) Molecular cloning and complete sequence of prion protein cDNA from mouse brain infected with the scrapie agent. Proc Natl Acad Sci 83:6372-6376
80. Lowenstein DH, Butler DA, Westaway D, McKinley MP, DeArmond SJ, Prusiner SB (1990) Three hamster species with different scrapie incubation times and neuropathology encode distinct prion proteins. Mol Cell Biol 10:1153-1163
81. Masters CL, Gajdusek DC, Gibbs CJ Jr (1981) Creutzfeldt-Jakob disease virus isolations from the Gerstmann-Sträussler syndrome: with an analysis of the various forms of amyloid plaque deposition in the virus-induced spongiform encephalopathies. Brain 104:559-588
82. Masters CL, Harris JO, Gajdusek DC, Gibbs CJ Jr, Bernoulli C, Asher DM (1979) Creutzfeldt-Jakob disease: patterns of worldwide occurrence and the significance of familial and sporadic clustering. Ann Neurol 5:177-188
83. McPherson, Shlichta P (1988) Heterogeneous and epitaxial nucleation of protein crystals on mineral surfaces. Science 239:385-387
84. Merz P, Sommerville RA, Wisniewski HM (1983) Abnormal fibrils in scrapie and senile dementia of the Alzheimer type. Virus non-conventionnels et affections du system nerveux central. Masson, Paris
85. Meyer RK, McKinley MP, Bowman KA, Braunfeld MB, Barry RA, Prusiner SB (1986) Separation and properties of cellular and scrapie prion proteins. Proc Natl Acad Sci USA 83:2310-2314
86. Miller DC (1988) Creutzfeldt-Jakob disease in histopathology technicians. N Engl J Med 318:853-854
87. Mitrová E (1990) Analytical epidemiology and risk factors of CJD. In: Court LA, Dormont D, Brown P, Kingsbury DT (eds) Unconventional virus diseases of the central nervous system. Commissariat á l'Energie Atomique, Service de Documentation, Paris
88. Neugut RH, Neugut AI, Kahana E, Stein Z, Alter M (1979) Creutzfeldt-Jakob disease: familial clustering among Libyan-born Israelis. Neurology 29:225-231
89. Nieto A, Goldfarb LG, Brown P, Wexler P, Chodosh HL, McCombie WR, et al. (1991) Mutation in codon 178 of amyloid precursor gene occurs in Creutzfeldt-Jakob disease families of diverse ethnic origins. Lancet, in press
90. Oesch B, Westaway D, Wälchli M, McKinley MP, Kent SBH, Aebersold R, et al. (1985) A cellular gene encodes scrapie PrP 27-30 protein. Cell 40:735-746
91. Owen F, Poulter M, Collinge J, Crow T (1990a) Codon 129 changes in the prion protein gene in Caucasians. Am J Hum Genetics 46:1215-1216
92. Owen F, Poulter M, Shah T, Collinge J, Lofthouse R, Baker H, et al. (1990c) An in-frame insertion in the prion protein gene in familial Creutzfeldt-Jakob disease. Mol Brain Res 273-276
93. Owen F, Poulter M, Lofthouse R, Collinge J, Crow TJ, Risby D, et al. (1989) Insertion in prion protein gene in familial Creutzfeldt-Jakob disease. Lancet i:51-52
94. Prusiner SB (1982) Novel proteinaceous infectious particles cause scrapie. Science 316:136-144

95. Prusiner SB, Gajdusek DC, Alpers MP (1982) Kuru with incubation periods exceeding two decades. Ann Neurol 12:1-9
96. Prusiner SB, Scott M, Foster D, Pan KM, Groth D, Miranda C, et al. (1990) Transgenetic studies implicate interactions between homologous PrP isoforms in scrapie prion replication. Cell 63:673-686
97. Roos R, Gajdusek DC, Gibbs CJ Jr (1973) The clinical characteristics of transmissible Creutzfeldt-Jakob disease. Brain 96:1-20
98. Safar J, Ceroni M, Gajdusek DC, Gibbs CJ Jr (1991) Differences in the membrane interaction of scrapie amyloid precursor proteins in normals and scrapie- or Creutzfeldt-Jakob disease-infected brains. J Infect Dis, in press
99. Safar J, Ceroni M, Piccardo P, Liberski PP, Miyazaki M, Gajdusek DC, Gibbs CJ Jr (1990a) Scrapie-associated precursor proteins: Antigenic relationship between species and immunocytochemical localization in normal, scrapie and Creutzfeldt-Jakob disease brains. Neurology 40:513-517
100. Safar J, Ceroni M, Piccardo P, Liberski PP, Miyazaki M, Gajdusek DC, Gibbs CJ Jr (1990b) Subcellular distribution and physicochemical properties of scrapie precursor protein and relationship with scrapie agent. Neurology 40:503-508
101. Safar J, Wang W, Paogett MP, Ceroni M, Piccardo P, Zopf D, Gajdusek DC, Gibbs CJ Jr (1990c) Molecular mass, biochemical composition and physicochemical behavior of the infectious form of the scrapie precursor protein monomer. Proc Natl Acad Sci USA 87:6373-6377
102. Sasaki H, Yoshioka N, Takagi Y, Sakaki Y (1985) Structure of the chromosomal gene for human serum prealbumin. Gene 37:191-197
103. Scott M, Foster D, Miranda CA, Serban D, Coufal F, Wälchli M, et al. (1989) Transgenic mice expressing hamster prion protein produce species-specific scrapie infectivity and amyloid plaques. Cell 59:847-857
104. Seitelberger F (1981) Spinocerebellar ataxia with dementia and plaquelike deposits (Sträussler's disease). Handbook of clinical neurology. North-Holland, Amsterdam
105. Sitwell L, Lach B, Atack E, Atack D, Isukawa (1988) Creutzfeldt-Jakob disease in histopathology technicians. N Engl J Med 318:854
106. Tateishi J, Sato Y, Ohta M (1983) Creutzfeldt-Jakob disease in humans and laboratory animals. In: Zimmermann HM (ed) Progress in neuropathology, vol. 5. Raven, New York, pp 195-221
107. Taylor DM, McConnell I (1988) Autoclaving does not decontaminate formol-fixed scrapie tissues. Lancet:1463-1464
108. Tsuzuki T, Mita S, Maede S, Araki S, Shimada KJ (1985) Structure of the human prealbumin gene. Biol Chem 260:1224-1227
109. Wakasugi S, Inomoto T, Yi S, Naito M, Ushira M, Iwanaga T, et al. (1988) A potential animal model for familial amyloidotic polyneuropathy through introduction of human mutant transthyretin gene into mice. In: Takashi I, Shakuro A, Fumike O, Shozo K, Eiro T (eds) Amyloid and amyloidoses. Plenum, New York, pp 393-398
110. Westaway D, Prusiner SB (1986) Conservation of the cellular gene encoding the scrapie prion protein. Nucl Acids Res 14:2035-2044
111. Westaway D, Goodman PA, Miranda CA, McKinley MP, Carlson GA, Prusiner SB (1987) Distinct prion proteins in short and long scrapie incubation period mice. Cell 51:651-662
112. Wilesmith JW, Wells GAH, Cranwell MP, Ryan JBM (1988) Bovine spongiform encephalopathy: epidemiological studies. Vet Rec 123:638-644

113. Yi S, Tadahashi K, Wakasugi S, Yamamura K, Araki S (1990) Pathological similarity to human familial amyloidotic polyneuropathy (FAP) type 1 in transgenic mice carrying the human mutant transthyretin gene. Abstract No.07/5 in Program and Abstracts of the VIth international symposium on amyloidosis, Oslo, 5-9 Aug 1990, p 58

114. Zu Rhein GM (1969) Electronmicroscopy of kuru. In Discussion on kuru. In: Burdzy K, Kallós P (eds) Pathogenesis and Etiology of Demyelinating Diseases. Additamentum to International Archives of Allergy and Applied Immunology, vol 36. Karger, Basel, pp 604-606

How Vaccines Work: Different Patterns for Different Diseases

J. Salk

The Salk Institute, P.O. Box 85800, San Diego, CA 92186-5800, USA

Different patterns for success have emerged in the course of my experience in the development of a science of vaccinology. The state of the art of the science 50 years ago led to the belief that for successful immunization against a virus disease it would be necessary to experience infection, albeit with an attenuated form of the virus, as in the case of smallpox, rabies and yellow fever [1]. As a medical student, I was struck by the apparent paradox that we could use chemically treated toxins but not chemically treated viruses for effective and durable immunity. In one lecture we were told that it was possible to immunize against diphtheria and tetanus by use of their toxins rendered harmless by chemical treatment. In the very next lecture we were told it was necessary, for inducing immunity to virus diseases, to use an attenuated live virus vaccine capable of causing infection and that such immunity could not be induced by a so-called "killed" or non-replicating virus vaccine. To me, it seemed that both statements could not be true and that one must be false. I reasoned that since immunization against diphtheria and tetanus was possible using the chemical substance of the toxin as vaccine that it should be possible to do the same with viruses without requiring that the virus multiply within the host in order to induce immunity.

Influenza

When the opportunity presented itself in 1939, while still a medical student, to work with influenza viruses, I subjected my hypothesis to test using ultraviolet radiation instead of chemical treatment to destroy virus infectivity and found that it was in fact possible to induce protective immunity in mice [2]. The key appeared in the dose-response pattern revealing the necessity for a sufficient quantity of virus to induce the desired effect. The significance of this pattern suggested that it would, therefore, not be necessary to require virus multiplication within the host if a sufficient quantity of virus in a noninfectious form could be administered.

As a result of the dose-response pattern using ultraviolet inactivated virus, a further study extended the pattern to include a chemically inactivated influenza virus using formalin that could confer immunity in human subjects as well as in mice [3]. A relationship was established between the level of neutralizing or hemagglutination-inhibiting antibody present in the serum and the degree of protection induced by vaccination [1].

The next pattern to emerge involved antigenic specificity as evidenced by the failure of the vaccines strains used successfully in 1943 [4] to protect against a new antigenic variant that appeared in 1947 [5]. The importance of antigenic drift and antigenic shift made it clear that the control of recurrent epidemics of influenza would require that vaccines include a sufficiently wide spectrum for antigenic coverage of appropriate specificity. Additionally, sufficiently high levels of antibody would be required which could be induced by an immunopotentiator in the form of incomplete Freund's adjuvant [6]. The use of a potent adjuvant would permit the inclusion of lower doses of many more strains to induce the desired effect [1].

Another pattern was observed in closed population groups in an institutional setting in which immunization of half the population protected the unimmunized as compared to a totally unimmunized group [7]. This pattern suggested that a herd effect was induced through a reduction in frequency of virus disseminators in the partially immunized population.

Thus, the patterns observed in studies on immunization against influenza suggested that a noninfectious virus vaccine could protect against infection and/or disease and that effectiveness depended upon the induction and maintenance of high enough levels of antibody of appropriate specificity. These studies also indicated the value of a potent immunological adjuvant in broadening the specificity and enhancing the level of the immune response.

Poliomyelitis

Similar patterns to those observed in studies on influenza also emerged in studies on polio virus immunization with a noninfectious virus vaccine. It soon became apparent, however, that persistence of immunity to paralysis did not depend upon the presence of detectable levels of antibody in the serum but rather on the presence of immunological memory which could readily be induced using a noninfectious polio virus vaccine [8]. Previously, it had been presumed that unless booster doses were administered or repeated exposure to natural infection would occur, antibody titers would decline with time and that upon subsequent exposure, infection and paralysis might ensue later in life [1, 9].

The patterns of the anamnestic response to immunization or to infection indicate that the intensity of immunological memory so induced is dose dependent [10]. Also, the intensity of anamnestic responsiveness develops over time after first exposure to antigen, as shown by the degree of responsiveness to a second dose when tested at intervals following the primary dose. This reaches a maximum after an interval of 6 months and appears to persist thereafter [11].

The presence of immunological memory even in the absence of detectable levels of antibody could be demonstrated by the characteristic anamnestic-type response measurable within a week following a challenge dose of vaccine [12]. Under circumstances of natural infection, this acceleration of antibody response has the effect of blocking virus spread in the blood stream from the primary site of infection in the intestine to the CNS and to the pharynx as well. By blocking CNS infection, paralysis is prevented in the individual and by blocking pharyngeal infection, the spread of virus to others is prevented, thus protecting the community as well.

In countries that have relied upon the exclusive use of an inactivated polio virus, vaccine has resulted in the disappearance of polio virus from their populations. Since humans are the only reservoir for polio viruses in nature, it would appear that eradication of polio virus from the human population is feasible using a noninfectious polio virus vaccine without the need for exposure to the experience of infection.

Since the persistence of immunity to paralytic polio requires merely the persistence of immunological memory, even a single dose of vaccine of sufficient potency can induce protection if an enhanced potency vaccine is administered, after the decline of maternal antibody, at about 6 months of age or later. However, if administered prior to 6 months of age, at a time when significant levels of maternal antibody may be present, a subsequent dose after 6 months of age is required to ensure an effective immune response [11, 13].

By comparison with the patterns associated with effectiveness of an inactivated virus vaccine for influenza requiring significant levels of neutralizing antibody at the time of exposure, in the case of poliomyelitis all that is required is the presence of immunological memory to protect against paralysis even though infection of the intestinal tract may occur. Thus, the patterns of the way vaccines work in these two virus diseases differ from each other and differ from what will by required for an effective human immunodeficiency virus (HIV) vaccine [14, 15].

HIV and Acquired Immunodeficiency Syndrome

By contrast with influenza and polio, in which the pattern of infection is that induced by cell-free viruses, HIV infection is usually transmitted by cell-associated virus in the form of HIV-infected peripheral blood lymphocytes, as in semen or vaginal secretions, by blood transfusion, or as a result of intravenous drug use. If transmitted as cell-free virus as may be contained in purified blood products, the virus can rapidly become cell associated [15].

Thus, to prevent the establishment of HIV infection and/or disease, it will be necessary for an HIV vaccine to induce the appropriate types of immune responses which are different from those required to protect against infection and/or disease by such cell-free viruses as influenza or poliomyelitis. While induction of type- and strain-specific neutralizing antibody is the necessary and sufficient requirement for protection against infection and/or disease in the case of influenza and poliomyelitis, it is unlikely that this will be sufficient for HIV and acquired immunodeficiency syndrome (AIDS). The extent to which neutralizing antibody is required for effective immunization against HIV remains to be seen.

From studies in the simian model [16], it is now clear that an inactivated simian immunodeficiency virus (SIV) vaccine, incorporating a cell-mediated immunity (CMI) adjuvant, can induce measurable protection, the duration of which remains to be determined. From the patterns observed thus far, it appears that, to establish and maintain a state of immunity to HIV infection, induction and persistence of cell-mediated immunological memory will be required to permit recognition and destruction of virus-expressing cells, as well as humoral immunological memory to inhibit virus infection.

Studies are underway to explore the prospects of forestalling the progression of HIV infection and the deterioration of the immune system by vaccination after infection but before oneset of disease (AIDS). The purpose would be to reduce the virus burden, or to maintain it at a low level by destroying virus-expressing cells, and to control whatever autoimmune phenomenon may be operative in CD-4 cell destruction [14]. The extent to which this will be possible will depend upon the capacity to induce and maintain an adequate level of CMI memory against HIV antigens using an appropriately prepared and appropriately administered immunogen together with an appropriate adjuvant.

Summary

The foregoing is intended to reveal how vaccines work in different diseases and how the approach to the development of an HIV vaccine must differ from that for vaccines against influenza or poliomyelitis or other virus diseases [14, 15]. It is evident, therefore, that appropriate strategies for the development and use of vaccines for particular infectious diseases must take into consideration the special features of each. This also applies to the prospects for the control of infectious or parasitic diseases in general and to the control of autoimmune and neoplastic diseases by the use of vaccines for prophylaxis and/or therapy.

Thus, for the development of effective vaccines for different diseases amenable to control by vaccination, there is the need to take account of the quantitative and qualitative requirements for inducing the appropriate immunologic response, and the appropriate antigenic elements required for each. By carrying out field studies to determine what these may be, the critically essential elements required for effective immunization against influenza and poliomyelitis were established. Based on such knowledge, it is now possible to improve upon the methods presently in use for immunization against influenza and poliomyelitis and with an economy of means. In due course, this will also become known for HIV/AIDS, when human field studies can be undertaken and the results of such experiences will become known.

References

1. Salk J, Salk D (1977) Control of influenza and poliomyelitis with killed virus vaccines. Science 195:834-847
2. Salk J, Lavin GI, Francis T Jr (1940) The antigenic potency of epidemic influenza virus following inactivation by ultraviolet radiation. J Exp Med 72/6:729-745
3. Salk J, Pearson H E, Brown PN, Francis TJr (1944) Protective effect of vaccination against induced influenza B. Proc Soc Exp Biol Med 55:106-107
4. Salk J, Menke W J Jr, Francis T Jr (1945) A clinical epidemiological and immunological evaluation of vaccination against epidemic influenza. Am J Hyg 42/1:57-93
5. Francis T Jr, Salk J, Quilligan J J Jr (1947) Experience with vaccination against influenza in the spring of 1947. Am J Public Health 37/8:1013-1016
6. Salk J, Contakos M, Laurent A M, Sorensen M, Rapalski A J, Simmons I H, Sandberg H (1953) Use of adjuvants in studies on influenza immunization. 3. Degree of persistence of antibody in human subjects two years after vaccination. JAMA 151/14:1169-1175
7. Salk J, Pearson H E, Brown P N, Smyth C J, Francis T Jr (1945) Immunization against influenza with observations during an epidemic of influenza A one year after vaccination. Am J Hyg 42/3:307-322
8. Salk J (1955) A concept of the mechanism of immunity for preventing paralysis in poliomyelitis. Ann N Y Acad Sci 61:1023-1036
9. Salk J (1959) Preconceptions about vaccination against paralytic poliomyelitis. Ann Intern Med 50/4:843-861
10. Salk J, van Wezel A L, Stoeckel P, van Steenis G, Schlumberger M, Meyran M, Rey J-L, Lapinleimu K, Bottiger M, Cohen H (1981) Theoretical and

practical considerations in the application of killed poliovirus vaccine for the control of paralytic poliomyelitis. Dev Biol Stand 47:181-198
11. Salk J (1984) One-dose immunization against paralytic poliomyelitis using a noninfectious vaccine. Rev Infect Dis 6 [Suppl 2]: S444-450
12. Swartz T A, Handsher R, Stoeckel P, Drucker J, Caudrelier P, van Wezel A L, Cohen H, Salk D, Salk J (1989) Immunologic memory induced at birth by immunization with inactivated polio vaccine in a reduced schedule. Eur J Epidemiol 5/2:143-145
13. Salk J (1990) Are booster doses of poliovirus vaccine necessary? Vaccine 8/5:419-420
14. Salk J (1987) Prospects for the control of AIDS through immunization of seropositive individuals. Nature 327:473-476
15. Salk J, Gersten M J (1990) Prophylactic and therapeutic immunization against AIDS: theoretical considerations underlying a modified non-infectious whole virus approach. In: Putney S, Bolognesi D (eds) AIDS vaccine: research and clinical trials. Dekker, New York, pp 265-275
16. Gardner M B, Stott J (1990) Progress in the development of simian immunodeficiency virus vaccines: a review. AIDS 4 [Suppl. 1]: S137-S141

The Future of Immunological Drugs

H.G. Schwick

Behringwerke AG, D-3550 Marburg (F.R.G.)

This year in November it will be 100 years since Behring and Kitasato published their work about achieving diphtheria immunity and tetanus immunity in animals in the *Deutsche Medizinische Wochenschrift* [1]. This was the beginning of the development and production of immunological drugs, which until today consist mainly of drugs containing antibodies and vaccines. Already at that time, very impressive figures existed in connection with vaccination against smallpox showing the convincing prophylactic efficacy of vaccine. The first statistical statement about the effectiveness of vaccination could be made more than 100 years ago after the Franco-Prussian war of 1870-1871. In the unvaccinated French army there were 125 000 cases of smallpox, and 23 000 of those infected died. The vaccinated German army on the other hand recorded only 495 deaths from smallpox. As a result of this experience, the public demanded obligatory smallpox vaccination, and support for this demand came from numerous life-insurance companies which had lost 14% of their insured persons from smallpox.

In April 1874, the Federal vaccination law was proclaimed in Germany. It was a model for other countries, and it played a decisive part in causing smallpox to disappear in Germany. The World Health Organization (WHO) could report a few years ago that because of systematic vaccination, the smallpox virus as a source of infection no longer existed. A great success for immunological prophylaxis. Other successful vaccinations followed. For example, about 30 years ago, oral vaccination against poliomyelitis was introduced and was a striking success. Furthermore, the efficacy of immunological prophylaxis by vaccination is shown by the protective immunization against measles.

I could provide many more such examples of the efficacy of vaccinations. Furthermore, in the last few years, cost-benefit analyses of vaccination against various diseases have been carried out by well-known institutes and these clearly show there

Table 1. Available vaccines

Type of infection	
Bacterial	Viral
Cholera	Hepatitis B
Diphteria	Influenza virus A + B
Hemophilus meningitis	Japanese encephalitis
Meningococcal meningitis	Measles
Pertussis	Mumps
Plague	Poliomyelitis
Pneumococcal pneumonia	Rabies
Tetanus	Rubella
Tuberculosis	(Smallpox)
Typhoid/paratyphoid fever	Tick-borne encephalitis
	Varicella
	Yellow fever

is also economic benefit to be gained from vaccination campaigns. Vaccines are the cheapest and at the same time, the most efficient drugs [2]. There are no other drugs which, after being administered once or twice, grant protection against diseases for many years or even for life. Today, more than 20 vaccines are available against bacterial and viral infectious diseases (Table 1).

With the exception of antivenom sera, immunoglobulins from animal sera play practically no part in prophylaxis and therapy today, since a little more than 40 years ago, the development and production of human immunoglobulins was started in connection with human plasma fractionation. Today we have at our disposal a large arsenal of various polyspecific and special immunoglobulins, which have a particularly high antititer against certain infectious agents. Further immunoglobulins are being tested in clinical trials (Table 2).

The immunological drugs available today, i.e., the vaccines and immunoglobulins described briefly above, are based mainly on the immunological knowledge gained from the work of Koch, Behring, Kitasato, and Ehrlich at the turn of the twentieth century. Are we about to reach a turning point in the development of new immunological drugs? My answer is: Yes, I believe we are! When we look at the major events in the history of immunology of the last 20 years, we realize that there was an enormous increase in knowledge about the immune system. During this period, new techniques became available, such as the

Table 2. Human immunoglobulins

Specificity	Available for clinical use	In clinical trials
Bacterial infection	Diphteria *Pseudomonas aeruginosa* Tetanus	*Escherichia coli* J5 Hemophilus influenza B Klebsiella *Staphylococcus aureus*
Viral infection	Cytomegalovirus Hepatitis A Hepatitis B Measles Rabies Rubella Tick-borne encephalitis Varicella	
Polyspecific	IgG enzymatically treated IgG-7S IgG-7S chemically modified IgM	
Others	Anti-Rh_0(D)-Ig	

Ig, immunoglobulin

synthesis of carbohydrates and peptides, the hybridoma technique and genetic engineering. These methods, together with the new scientific knowledge of the immune system, were decisive for the development of new immunological drugs.

If we regard the immunological system today as an organ in which various cells with different functions interact, e.g., soluble proteins like antibodies and immune factors such as interleukins and cytokines, numerous possibilities arise for the development of immunological drugs. Beyond the treatment of immunodeficiency diseases and the prophylaxis and treatment of infectious diseases, immunological drugs will have great importance for transplantations, various autoimmune diseases, allergies, and even for the treatment of tumors.

The working mechanism will be the substitution, stimulation, or regulation of the immunological system or its mitigation in transplantations, autoimmune diseases, and allergies. Many laboratories worldwide have started working on the development of monoclonal antibody preparations to be used in prophylaxis and therapy. These can be regarded as the third generation of immunological drugs, if we regard the antisera from animals as the first and immunoglobulins isolated from human plasma as the second. Monoclonal antibodies are highly specific, and today it is possible to produce them in large quantities and in a homogeneous and easily reproducible way. Animal or human blood donors are not required for their production. Most of the monoclonal antibodies available today for experimental examinations and first clinical trials are antibodies with animal, mostly murine, specifity. Most recently, methods have been developed which either allow the humanization of murine monoclonal antibodies or even the construction of human monoclonal antibodies. The fundamentals of the gene technology of antibodies, as we now understand them, make the humanization of monoclonal antibodies basically possible, and, furthermore, allow the construction of bispecific antigen-binding antibody molecules of different size.

Figure 1 shows the structures of various monoclonal antibodies. In antibody engineering, on the basis of these structures, an attempt is made to develop monoclonal antibodies against specific infectious agents. However, antibodies against cell receptors and against immune factors are also gaining more and more clinical interest as immunoregulatory drugs which might possibly become important in the treatment of autoimmune diseases. The clinical trials of a murine monoclonal antibody against the αß T-cell receptor in organ and bone marrow transplantations are proceeding successfully. Promising preclinical data exist.

Since the development of technology to produce monoclonal antibodies, their usage for the antigen-specific therapy of tumor diseases has been discussed. There are different starting points. Provided that they have the specifity and that they bind in a sufficient quantity to the tumor, monoclonal antibodies could directly hamper tumor function. They could destroy the tumor via the initiation of immunological antibody-dependent effects mechanisms such as complement activation or antibody-dependent cytotoxicity; they could also have antitumoral effect as vectors of cytostatics, isotopes, or toxins.

We know today that the localization of suitable monoclonal antibodies in tumor patients is sufficient for the immunoscintigraphical diagnosis of tumors. About 0.01% of the intravenously injected monoclonal antibody is to be found on 1 g

Table 3. Diseases that are potentially preventable by vaccines

Type of infection		
Bacterial	Parasitic	Viral
Chlamydia infections	Amoebic diarrhea	AIDS
Escherichia coli diarrhea	Chagas' disease	Dengue
Gonorrhea	Leishmaniasis	Hepatitis A
Lepra	Malaria	Hepatitis C
Meningococcal B meningitis	Nematodal infection	Herpesvirus
Pseudomonas infections	Schistosomiasis	Rotavirus
		Respiratory syncytial virus

AIDS, acquired immunodeficiency syndrome

of tumor. This concentration is about ten times too low to destroy the tumor as a conjugate with cytostatics, isotopes, or toxins, but it is astonishing that the monoclonal antibody localized on the tumor is reduced by the body to a far lesser degree than the remaining over 99% of monoclonal antibodies in the normal tissue. The reasons for this are still unknown, but the quantity of the continuous presence of the monoclonal antibody on the tumor in comparison with the normal tissue basically offers the possibility of a tumor therapy that is divided into two phases. After the monoclonal antibody has been injected and that portion of it which is localized in the normal tissue metabolized in the first phase, the remaining portion, which is now specifically localized on the tumor, can be used in a second phase, namely, cytotoxically and with or without an amplification system. Such cytotoxic effects can be, for instance, obtained by nontoxic prodrugs, which are split into a cytostatic by enzymes connected to the antibody on the tumor. A second possible amplification system is the induction of anti-idiotypic antibodies in the tumor host, and a third possibility is the use of those specific monoclonal antibody preparations which will effect a specific binding of the cytotoxic principle to the tumor after the metabolizing phase.

What new developments in vaccines are in sight? Table 3 shows a list of those diseases which are caused by bacteria, parasites, and viruses, diseases which could most probably be prevented by vaccination prophylaxis. Work is currently in progress in various laboratories to develop vaccines against practically all the infectious diseases listed here. Some of these vaccines are already in clinical trial. The development of vaccines against parasitic infections is a particularly great challenge to immunological research. The number of human beings exposed to parasitic infections is enormous, and the number of people who fall ill with malaria for the first time each year exceeds the number of people who fall ill with the various other virus diseases by a factor of 100. When we think of recent WHO and UNICEF statistics on infant mortality in the world, it is clear that more than one-fifth of those children die of malaria. The prerequisites for a malaria vaccine are, however, more favorable today than they were only a few years ago. Work on the development of a vaccine against malaria is being carried out with great intensity at present, so one can expect to have a vaccine in a clinical trial within a foreseeable time.

There are several new approaches to the immunological prophylaxis of infectious diseases such as recombinant life vaccines, polysaccharide-protein conjugate vaccines, recombinant epitope carrier vaccines, recombinant toxoids, and, last but not least, new formulations for oral and parenteral immunization. Recombinant life vaccines which have genetically engineered antigens tied to an nonpathogenic vector such as adenovirus could, in particular, have advantages such as long-term efficiency and efficacy, and, furthermore, these vaccines could be especially advantageous for large vaccination campaigns.

For a long time now, great efforts have been made to develop vaccines for the prophylaxis and immunotherapy of tumors. Recently, this work has been intensified in many laboratories, and, indeed, hepatitis vaccines are providing protection against hepatitis carcinoma. In preclinical trials, there are, for example, a human papillomavirus vaccine for the prophylaxis of cervical, rectal, and bronchial carcinoma, an Epstein-Barr vaccine, and a human T-cell lymphoma virus and human immunodeficiency virus vaccine. For the active antigen-specific immunotherapy of tumors, there are presently some vaccines in preclinical and clinical trials. These are either preparations of tumor cells which were modified by viruses, bacillus Calmette-Guérin, or neuraminidase, or isolated tumor-associated antigens and tumor virus antigens (Table 4).

For the sake of completeness, it should be mentioned that work is progressing in the production of vaccines for fertility regulation in the human and veterinary fields. For many years

Table 4. Active antigen-specific immunotherapy of tumors

Immunogen	Modification	Tumor	Preclinical	Clinical
Tumor cell (authochtoneous)	NDV	Breast carcinoma		+
Tumor cell (authochtoneous)	BCG	Colon carcinoma		+
Tumor cell (allogenic cell line)	VCN	Colon carcinoma		+
Isolated TAA				
GD_2, GD_3	Carrier	Melanoma	+	+
PEM	Carrier	Breast and ovary carcinoma	+	
GIT mucin	Carrier	Gastrointestinal carcinoma	+	
Tumor virus antigens	Carrier	Virus associated tumors	+	
	Vaccinia			

NDV, Newcastle disease virus; BCG, bacillus Calmette-Guérin; VCN, vitriocholerae neuraminidase; TAA, tumor-associated antigens; GD, gangliosid; PEM, polymorphic epithelial mucin; GIT, gastrointestinal tract

now, a task force group has existed within the WHO which is composed of scientists who are concerned with immunocontraception by vaccination. The main antigens which are examined here are gonadotropin-releasing hormone, luteinizing hormone, follicle-stimulating hormone, chorionic gonadotropin, ovum antigens, and sperm antigens such as LDH-C4.

After this survey of antibody and vaccine development I want to proceed to a third group of immunological drugs, which I shall call immune factors. These immune factors are of extreme biological significance, since, on the one hand, they are responsible for cell growth and reproduction and cell division and renewal, and, on the other hand, they can cause cell maturation and differentiation. Here, the change in the functional status of a cell is induced by a change in the surface antigen pattern, since maturation is accompanied by the expression of new functional surface structures (mainly receptors). How can we form a picture of these cell growth and differentiation factors? What is their effective mechanism, and how do they interact? There is apparently a unified scheme, in accordance with which a smaller pool of uncommitted stem cells produces, via many stages of

Table 5. Recombinant human growth and maturation factors of the hemopoetic and immune system

Cytokines (Interleukins)	Aminoacids (*n*)	Native molecular weight (KDa)
IL-1α	159	17
IL-1ß	153	17
IL-2	133	14-16
IL-3 (multi-CSF)	133	15
IL-4	129	20
IL-5	112	$(25)_2$
IL-6	184	22-29
IL-7	129	25
IL-8	99	10
IFNγ	143	25
TNFα	157	17
TNFß	169-171	20-25
GM-CSF	127	22
G-CSF	174-177	20
M-CSF	223	$(43\text{-}45)_2$
	145	$(20\text{-}25)_2$
EPO	166	34

IL, interleukin; CSF, colony stimulating factor; IFN, interferon; TNF, tumor necrosis factor; G, granulocyte; M, macrophage; EPO, erythropoetin

maturation, all the cells of the blood, including all the immune cells, as well as the platelets and the erythrocytes. Growth and maturation are largely subject to stimulation and control by growth and differentiation factors, which are effective either in earlier maturation stages (such as hemopoietin-1/interleukin-1, interleukin-3, and interleukin-6) or in later phases of maturation, e.g., granulocyte/macrophage-, macrophage-, and granulocyte-colony-stimulating factors or erythropoietin. The exact mechanism by which these individual factors are effective can be ascertained only when the substances are available in a uniform and pure state, which is possible by means of genetic engineering. A large number of these factors have been cloned in recent times with the aid of modern molecular biological techniques and can be produced in adequate quantities (Table 5).

Meanwhile the enormous physiological importance of these immune factors or cytokines as hormonal regulators of the growth and maturation of cells in our organism has been recognized. With them, new indications have arisen in areas in which we had

Table 6. Clinical uses of cytokines

Factor	Indication for uses
EPO	Anemia, renal failure, blood transfusion
GM-CSF	Leukopenia (radio-, chemotherapy), infection, aplastic anemia, BMT
G-CSF	Leukopenia, agranulocytosis, BMT, infection
IL-3	Leukopenia, thrombocytopenia, bone marrow restoration, BMT
IL-1	Radioprotection, wound healing, adjuvant
IL-2	Adoptive immunotherapy, killer cell activation, (tumor therapy)
IL-4	Tumor therapy, infectious diseases, immunoglobulin synthesis
IFNγ	Infectious diseases, rheumatoid arthritis

EPO, erythropoetin; G, granulocyte; M, macrophage; CSF, colony stimulating factor; IL, interleukin; IFN, interferon; BMT, bone marrow transplantation

previously no treatment available. Particularly for blood cell growth hormones, interleukins, and colony-stimulating factors (including the erythropoietin), large fields of clinical application can be recognized mainly in the context of disturbances and damages to hematopoiesis and the immune system, and in the treatment of infections and tumors (Table 6).

Some human interleukins and colony-stimulating factors are already available for clinical application, others are in an advanced stage of clinical trials.

Furthermore, it became apparent that the combination of these immune factors can lead to different effects on the maturation of cell growth. Granulocyte/macrophage stimulating factor, which alone has no influence on the platelets, influences together with interleukin-3 or even with erythropoietin the growth of blood platelets.

Recently, it has become possible to characterize and make biotechnologically accessible protein structures which serve as cell surface-linked specific receptor molecules for the respective cytokines and as a receiver of signals, and which initiate the activation of the target cells via carriers. Comparatively new is the discovery that physiologically existing, soluble, nonmembrane-linked forms of these receptor molecules cause the opposite reaction and neutralize cytokines before they reach the cell surface. Soluble receptors are to be understood as antagonists

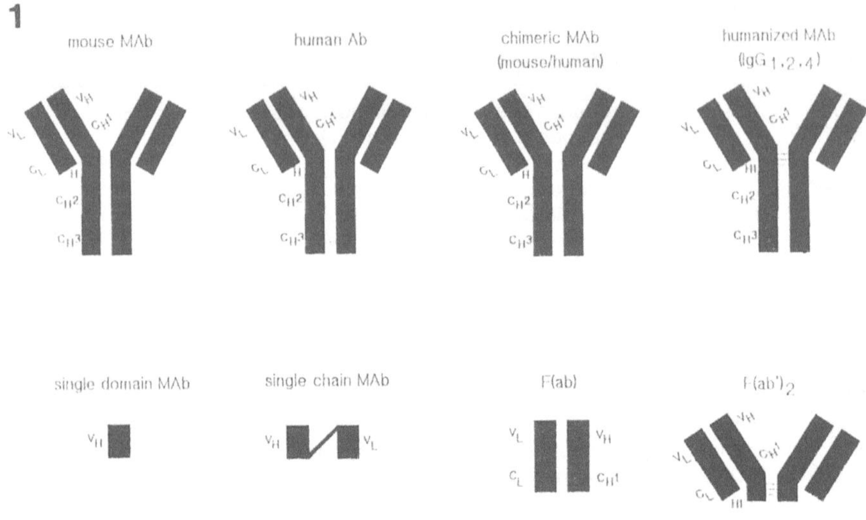

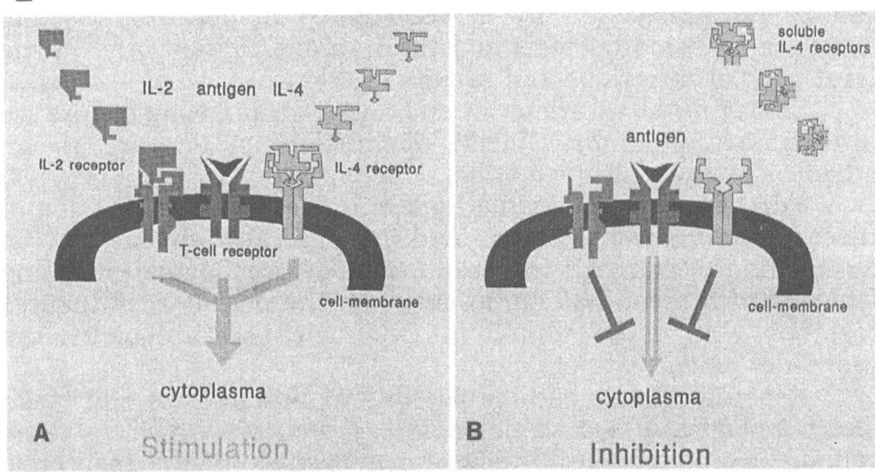

Fig. 1. Structure of various monoclonal antibodies. **Fig. 2.** Immunoregulation by cytokines and cytokine receptors. *a* T lymphocyte activation by binding of ligands (antigen plus cytokines). Therapeutic uses include immunostimulation and immunorestoration. *b* T lymphocyte inhibition by neutralization of cytokines. This is of therapeutic use in transplantation and autoimmune diseases

of the respective cytokines as well as in their constant cell membrane form. Medical fields in which interleukin receptors can be applied include immunosuppression, i.e., in transplantations and autoimmune diseases, the inhibition of chronic inflammable processes, cell proliferation inhibition, and partly antiallergenic processes. Figure 2 shows a schematic summary of the mechanisms of the cytokines and the cytokine receptors.

As already mentioned, antigens and interleukins bind via very specific cell receptors and the cell membrane to the cell and pass signals to it which will lead to its stimulation. Soluble, noncell-linked interleukin receptors bind the interleukin so that it will not reach its cell receptor. The result is the inhibition of cell division.

To summarize: With these new immunological drugs we will arrive at new and better concepts for the treatment of diseases of the immune system. Michael Arnold recently said [3] in connection with the new possibilities:

"It is hardly foreseeable at present what significance for therapy the receptor model will attain with which the differentiation of tissues and the functional control of cells can be explained. This might result in the possibility of a qualitatively completely new therapy of systemic diseases. One could succeed in modulating purposefully in the desired way the function of a cell or of complete organ systems. So in the end it would be realized in a modern form what since the classical experiments of Ehrlich has remained an important principle of every rational therapy: To develop drugs which act as specifically as possible on the target object. With Ehrlich these still were the agents which invaded the organism from outside to which the substance should be connected. Today these are the dysfunctional cells which should be reached and influenced in such a way that the function will be normalized."

Concluding remarks

I believe that we are now only at the beginning of a clinical evaluation of the new immunological drugs. The further problem of developing these substances lies not so much in the production of sufficient quantities, since that is becoming easier and easier thanks to molecular biology. The problem lies rather in designing and carrying out clinical studies in which these new factors are used alone and in combination with, e.g., chemotherapeutic substances. In this connection I like to close by citing the late Sir Ernest Chain, who together with Fleming discovered and

developed penicillin. He said at a meeting in Marburg in 1975:
"It is important to know that in bacterial chemotherapy, immune strength and direct antibacterial effect are complementary. It is therefore of great importance in spite of the successes and in the interest of antibacterial chemotherapy to enlarge the knowledge about specific and unspecific immune strengths and to look for methods to intensify them."
I believe the research on immunological drugs is well on the way to achieving this.

References

1. Behring E, Kitasato S (1890) Dtsch Med Wochenschr 16/49:1
2. Hireman AR, Jordan WS (1983) Public Health Reports 98:439
3. Arnold M (1988) Die medizinische Notwendigkeit neuer Behandlungskonzepte. MPS-Symposium, Bonn, FRG, 21 April 1988

Vaccines in Perspective: Human Hepatitis B Vaccines, the First Subunit and Recombinant Viral Vaccines

R. Hilleman

Merck Institute für Therapeutic Research, Merck Sharp & Dohme Research Laboratories, West Point, Pennsylvania 19486, USA

Historical Development of Viral Vaccines

The history of viral vaccines [1] that began in 1798 with Jenner's cowpox vaccine can be divided into several epochs (Table 1). The early vaccines consisted of whole viruses that were either live attenuated or killed. Vaccines developed during the first 150 years were made from the infected organs or tissues of animals and were principally those against smallpox, rabies, yellow fever, and viral influenza. Enders' cell culture renaissance revolutionized the technology for vaccines during the period from 1950 to 1980 and made possible vaccines against poliovirus, adenovirus, measles, mumps, rubella, varicella, and a hepatitis A prototype [2-4].

The decade of the 1980's saw the fruition of work begun in 1968 to develop a hepatitis B virus vaccine, which represents the first subunit viral vaccine [5,6]. It was followed by the first recombinant-produced viral vaccine, also against hepatitis B [7,8]. Hepatitis B virus cannot be grown in cell culture or in animals other than chimpanzees, so that these two hepatitis B vaccines bear the additional distinction of being the first vaccines against a nonpropagable virus.

Going to the future, it now appears possible to prepare vaccines that will be effective against nonpropagable and unidentified viral agents for which only cloned gene segments are available. Such technology may yield vaccines against hepatitis C [9,10] and hepatitis E [11] within the foreseeable future. In the more distant future, chemically synthesized antigens and highly polyvalent complexes of immunologic determinants of many

Table 1. Viral vaccines in historical perspective

Propagable agents		Nonpropagable agents		Propagable or nonpropagable agents	
Whole virus (live or killed)		Subunit virus			
1798-1949	1950-1980	1981	1986	Future	
Infected animal organs and tissues	Cell culture	Carrier plasma		Cloned segments	Synthetic polytopic
Smallpox Rabies Yellow fever Influenza[a]	Poliovirus Adenovirus Measles Mumps Rubella Varicella Hepatitis A[b]	Hepatitis B	Hepatitis B	Hepatitis C Hepatitis E ?	?

[a]Embryonated hens' eggs.
[b]To be licensed

different viruses should make it possible to prevent many different infections using a single complex or polytopic vaccine [12].

It should be noted that all viral vaccines developed to the present time act by stimulating production of specific antibodies that are known or are presumed to be neutralizing and that prevent viral infection. Some of the vaccines, principally those consisting of live virus, may also induce cytotoxic T cell immunity that is of importance in clearance of already infected cells.

Overall, the period from 1798 to the present was one of immense technical complexity as regards the development of each of a number of viral vaccines and the evaluations for their safety and efficacy, since the data bases were small, and guiding precedents were lacking. The period, however, was not given to exhaustive studies of pathogenesis of viral disease or to how a vaccine might function to avert these pathogenic mechanisms. It was the living example, instead, of the concept that "one can solve problems of immense importance without really understanding them." This simplicistic approach was very successful for the vaccines of the past, but it may not prove as successful for preventing infection with more complicated viruses such as those that cause acquired immunodeficiency syndrome (AIDS).

Hepatitis B

With this background, I should wish to devote the remainder of my lecture to hepatitis B and hepatitis B virus vaccines that were the start of a new era in vaccines, in pioneering, and in establishing the precedents for the first subunit viral vaccine and for the first recombinant-produced viral vaccine that may be utilized in developing new vaccines in the future.

Human Hepatitis B

Human hepatitis B virus infection causes liver disease, which affects people in most parts of the world. Though acute viral infection may lead to death, the more serious consequences are related to a carrier state that often causes chronic liver disease with cirrhosis and hepatocarcinoma 20 or 30 years later. There are about 300 million carriers of hepatitis B virus infection in the world population as of now, and of these, 75 to 100 million persons will die of liver cirrhosis and/or cancer. The danger of developing the carrier state with hepatitis B virus is greatest in early life and diminishes with increasing age. Up to 90% of babies born to carrier mothers may become carriers themselves. Administration of hepatitis B vaccine prevents the carrier state, even when given to the newborn and offers the means and a promise for eradicating hepatitis B from the world population.

Hepatitis B Vaccines

Plasma-Derived Vaccine. Hepatitis B virus cannot be propagated in vitro. This was the deterrent for many decades to any fruitful development of a vaccine. The discoveries by Prince [13,14] and by Blumberg [15] of hepatitis B surface antigen in the blood of carriers made it possible for subsequent development of a vaccine [5,6] using antigen purified from human carrier plasma. Work begun in 1968 led to clinical tests of a plasma-derived vaccine in 1975 [5,6] and licensure of this first hepatitis B vaccine in 1981.

Plasma from a human carrier of hepatitis B (Fig. 1) may contain large numbers of circular Dane virus particles plus large numbers of 22-nm diameter particles that are made of excess surface antigen of the virus. Purification and/or destruction of all possible life forms in the purified antigen was accomplished using the following series of steps:

- Centrifugation
  - Isopycnic banding
  - Rate zonal sedimentation
- Critical inactivation steps
  - Pepsin digestion
  - Urea denaturation and renaturation
  - Treatment with formaldehyde

The purified antigen (Fig. 2) was incorporated into an alum adjuvant. The vaccine was highly safe and effective in stimulating antibody in more than 95% of persons who received it. All persons who developed antibody were protected against infection and the carrier state. Hepatitis B vaccine, then, was the first vaccine against cirrhosis of the liver and human cancer.

Recombinant Yeast-Derived Vaccine. It soon became evident that a new source of antigen for preparing human hepatitis B vaccine should be sought, since (a) there is only a limited supply of suitable human carrier plasma; (b) the process for manufacture from human blood is slow and tedious; and (c) there are unfounded but real fears about the safety of human blood as a source of antigen for a vaccine. It became evident in 1975 that the new science of molecular biology had developed to the point of substituting microbial cells as the producer cell for making antigen in place of those of the infected human liver. The work with recombinant antigen culminated in the almost simultaneous licensure in West Germany and in the United States of the first recombinant hepatitis B vaccine in 1986 [7,8].

The double-stranded DNA genome (Fig. 3) of human hepatitis B virus has four open reading frames. The gene for the virus core encodes the internal viral polypeptides. The surface antigen gene consists of three segments called Pre-S1, Pre-S2, and S, each of which has its own initiation codon and a common terminator.

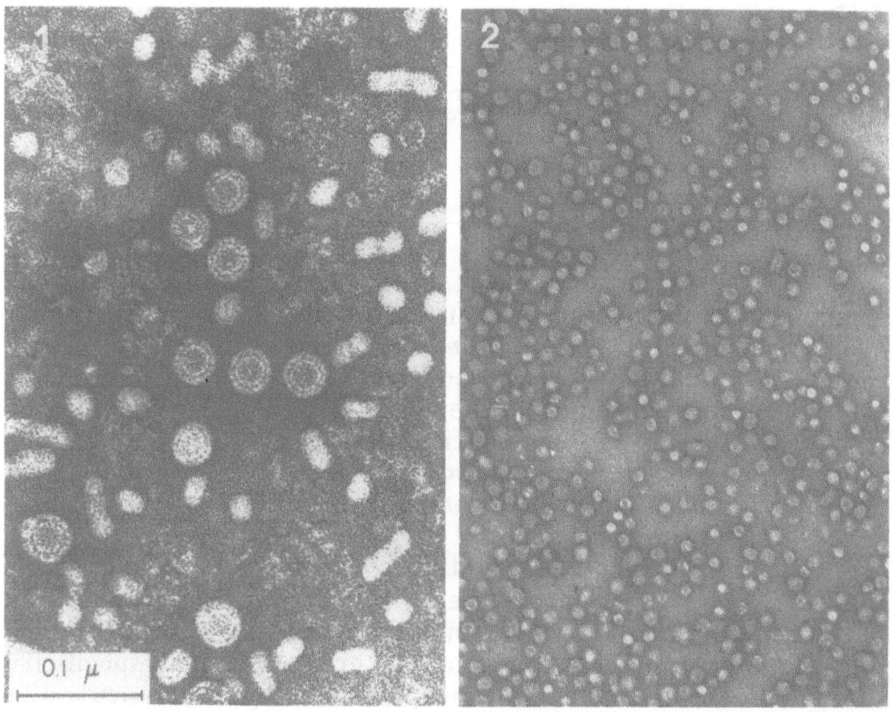

Fig. 1. Viral and surface antigen elements found in crude hepatitis B preparations in the plasma of a human hepatitis B carrier. **Fig. 2.** Hepatitis B vaccine, multiplication sign x 56 000

S antigen alone is included in the plasma-derived vaccine. In studies in collaboration with Valenzuela et al. [16], the S antigen of the viral DNA was excised from the double-stranded viral DNA, flanked by promoter and terminator sequences, and inserted into the ring structure of a plasmid vector (Fig. 4). Transfected into yeast, this plasmid faithfully encodes for production of surface antigen polypeptide that then can be readily purified from the yeast cell lysate.

Polypeptide Folding. The hepatitis B S antigen, as produced in yeast, is a linear sequence of 226 amino acids (Fig. 5). In its primary structure, the S antigen has no capacity to elicit antibodies relevant to protection against hepatitis B virus infection. As for other globular polypeptides, the primary chain

Table 2. Critical steps in achieving maximal immunizing potency of recombinant hepatitis B vaccine

Maximal yield of amount of antigen per amount of yeast protein (final purity of antigen >99%)
Thiocyanate treatment
Formaldehyde treatment
Alum-antigen co-precipitation

assumes secondary and tertiary structures, and folds into a globular form. In its folding, the S antigen develops at least one conformational *a* antigen epitope that is seen by B cells and that confers immunity against all hepatitis B virus subtypes. One or more immunodominant sequential epitopes are seen by T helper cells after endocytosis, degradation to short peptides, and presentation with major histocompatibility complex (MHC) class II receptor on the surface of antigen-presenting cells.

A computer model of hepatitis B S antigen [17] is shown in Figure 6. The largest part of the S antigen is hydrophobic and is buried in the lipid of the viral envelope; there is a short sequence that lies within the virus particle. The group *a* antigen specificity lies within the confines of two loops formed by amino acids 124 through 147 [18].

Immunizing potency, per unit amount, or weight of polypeptide antigen, depends upon the correctness of folding of the polypeptide to give a maximal epitope density and optimal three-dimensional or spatial configuration. This is achieved (Table 2) by appropriate antigen processing that (a) gives initial maximal purity (final purity of antigen >99%); (b) uses a thiocyanate treatment step to loosen and reform optimal disulfide bonds; (c) is stabilized by treatment with formaldehyde which is believed to build urea bridges between free amino groups; and (d) maximizes alum-antigen association through co-precipitation.

Surface Antigen Composition. The surface antigen particles isolated from human carrier plasma or prepared from

Fig. 3. Hepatitis B virus genome and defined antigens that are produced from it. **Fig. 4.** Construct of plasmid (pHBS56GAP347/33) used to produce hepatitis B surface antigen in yeast. **Fig. 5.** Forms of polypeptide epitopes. **Fig. 6.** Predicted secondary structure, hepatitis B S antigen. (From [17])

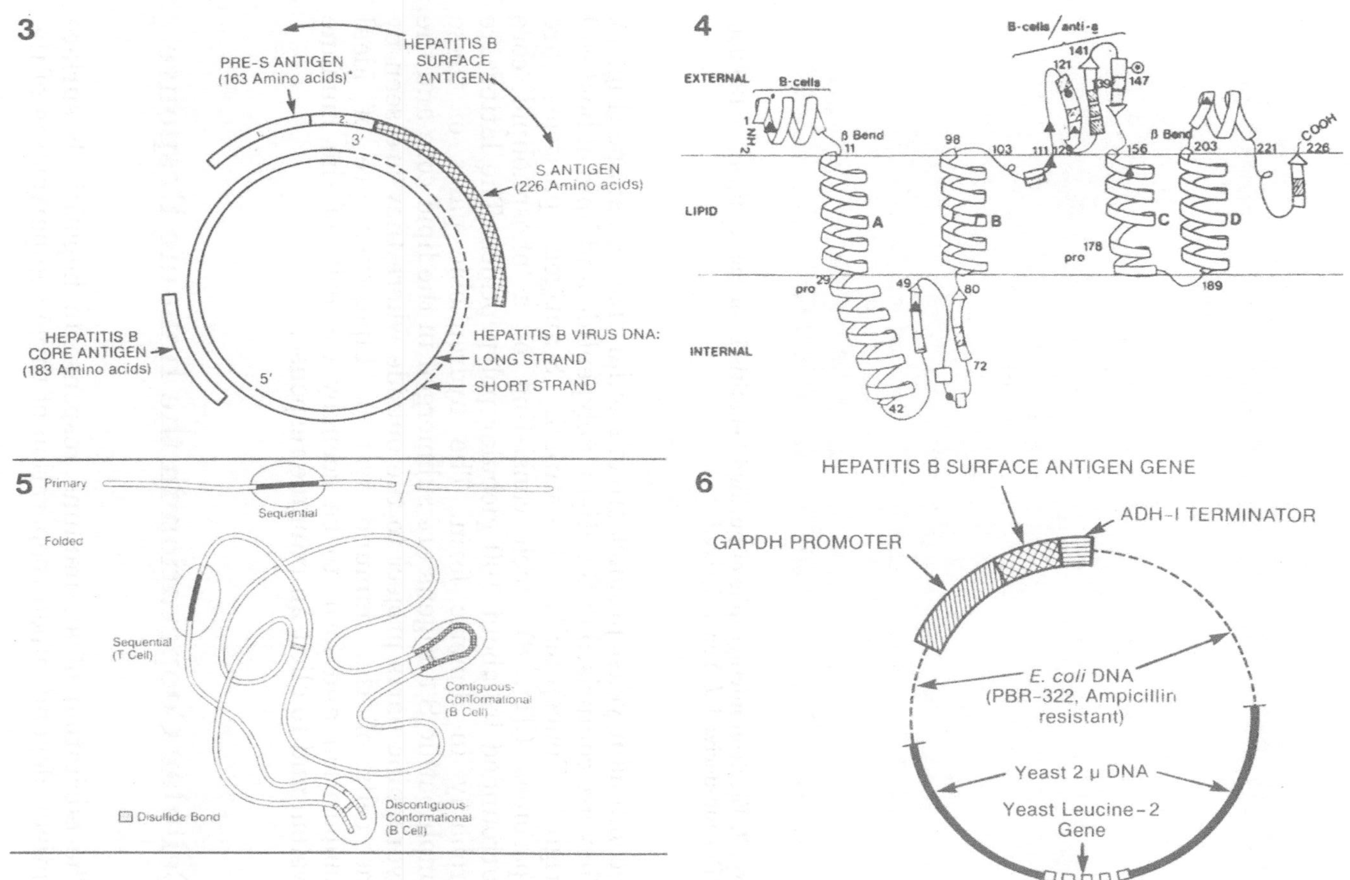
3
PRE-S ANTIGEN
(163 Amino acids)*
HEPATITIS B
SURFACE
ANTIGEN
S ANTIGEN
(226 Amino acids)
HEPATITIS B
CORE ANTIGEN
(183 Amino acids)
HEPATITIS B VIRUS DNA:
LONG STRAND
SHORT STRAND
3′
5′
4
EXTERNAL
LIPID
INTERNAL
B-cells
B-cells / anti-a
β Bend
NH2
COOH
A
B
C
D
pro
5
Primary
Sequential
Folded
Sequential
(T Cell)
Contiguous-
Conformational
(B Cell)
Discontiguous-
Conformational
(B Cell)
Disulfide Bond
6
HEPATITIS B SURFACE ANTIGEN GENE
ADH-I TERMINATOR
GAPDH PROMOTER
E. coli DNA
(PBR-322, Ampicillin
resistant)
Yeast 2 μ DNA
Yeast Leucine-2
Gene

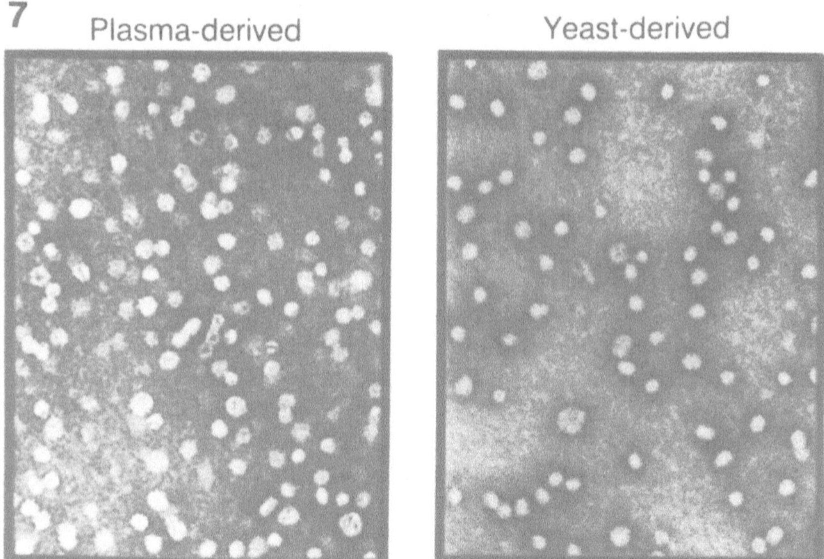

Fig. 7. Electron micrographs of purified hepatitis B surface antigen, x 130 000. *a* Plasma-derived. *b* Yeast-derived

recombinant yeast product are of similar size and shape (Fig. 7). They are comprised of S antigen polypeptide and lipid of host cell origin. Conceptually, hepatitis B S antigen particles are liposomes (Fig. 8), each consisting of a phospholipid core surrounded by about 100 globular polypeptides. The latter are probably in dimeric form. The hydrophobic parts of these amphipathic S antigens are submerged in the lipid core and the hydrophilic parts project to the outside where they are seen by the cells of the immune system. Liposomes are of ideal composition and form to be seen by the cells of the immune system and to elicit an immune response.

Cellular Cooperation in the Immune Response

The elicitation of an immune response to hepatitis B surface antigen depends upon cooperation of three compartments of the immune system, monocyte/macrophage antigen-presenting cells, T helper cells, and antibody-producing B cells (Fig. 9).

Hepatitis B antigen enters the internal body environment of the susceptible individual. Resting B cells of complementary specificity engage the conformational loop epitopes of the hepatitis B antigen and begin to differentiate and divide, but they need help, and this is provided by T helper lymphocytes. Cells of the macrophage/monocyte series engulf the hepatitis B antigen, break it down to small linear epitopes and present them on the cell surface in association with MHC class II receptor. Resting T helper cells of complementary specificities establish contact with the linear epitope on the antigen-presenting cells and receive lymphokine stimulation from them, primarily interleukin-1 (IL-1). The primed T helper cell now generates interleukin-2 (IL-2) and seeks to engage a corresponding B cell. The primed B cells, like the antigen-presenting macrophages, ingest whole antigen and process and present the same linear epitope on their cell surface in association with the MHC class II receptor. This is the signal to link the primed helper cell with the complementary primed B cell. This facilitates division, avidity maturation, and differentiation into antibody-producing plasma cells.

Not all persons are capable of responding to all antigens. A small proportion of human beings are incapable of response. This is due to the polymorphism of the MHC class II receptor pocket that, in most individuals, can accept and present most, but not all linear epitopes (Fig. 10). The epitope binding site of the MHC class II receptor [19], like that for MHC class I, is believed to consist of a groove or pocket that is made of flat sheets in its bed and α helices on each side which vary and confer a degree of selective immunologic capability for presenting a particular epitope. The degree to which the MHC class II receptors of the various MHC repertoire in a particular individual can accept and present an epitope determines the degree of T helper response.

Hepatitis B Vaccination

Immunogenicity. Hepatitis B virus vaccine [20] is given to human subjects in appropriate dose (antigen amount) and in a fixed

Table 3. Protective efficacy of recombinant hepatitis B vaccine in neonates born to carrier mothers

Principal investigator	Immunoglobulin	Location	Protective efficacy (%)
C.E. Stevens	Yes	U.S.A.	94
E.K. Yeoh	Yes	Hong Kong	94
K.M. Tin	No	Burma	92
Z.-H. Hu	No	China	75[a]

[a]Estimate at 6 months.

regimen that usually consists of two priming doses followed by a booster dose after a rest period of 6 months to 1 year (Fig. 11). Amounts of antigen adequate to achieve 100% or near 100% seroconversion of persons at a 10 mIU/ml or higher titer are used. Larger amounts of antigen will confer higher levels of antibody, but, for purposes of economy, only adequate doses are needed since the criterion for protection is an adequate antibody response; an excess of antibody confers no advantage.

The height of antibody response to antigen is highly age-related and describes a bell-shaped curve. Very young infants and elderly patients develop the least amount of antibody, while the maximal antibody response is in the teenage years. The vaccine dose is adjusted to accommodate these realities.

Protective Efficacy. Protective efficacy in vaccinated human beings is determined by measuring prevention of infection, prevention of illness, and prevention of the carrier state as evidenced by failure of appearance of circulating surface antigen on effective exposure to virus. Four clinical studies (Table 3) to measure protective efficacy were carried out with recombinant hepatitis B vaccine. In the studies, infants born to *e* antigen (highly infectious) positive mothers were immunized, starting within 12 h after birth. In two studies, passive hepatitis B immunoglobulin (Ig) was given to provide immediate protection; Ig was not given in the two other studies. Protective efficacy was calculated in terms of the amount of reduction in carrier state in the infants at 9 or 12 months of age compared with the historically established carrier state in unvaccinated infants.

In the studies in which passive antibody was given together with the first dose of vaccine, protective efficacy was 94%. In

Burma, where infection is acquired mainly by horizontal transmission after birth, the efficacy was 92% in the absence of antibody. In China, where infection is heavily perinatal, the efficacy was only 75%.

Future Considerations

Duration of immunity. Concern is presently being expressed as to how long immunity will last following vaccination, and whether there is or is not need for a late booster dose of vaccine at present or in the future. There are two views. The one calls for surveillance of antibody levels in vaccinated persons and reimmunization at a calculated time period or when the measured level of antibody declines to a prescribed low level. The other view is that immunity does not depend upon detectable circulating antibody, but, rather, upon a retained capability for lymphocytes to give an immediate anamnestic response upon encounter with the virus. The latter judgment accords with the majority view and with current World Health Organization and United States Public Health Service policy not to pursue routine reimmunization. However, there can be no harm for reimmunization for those persons whose resources permit it.

The lack of need, at this time, for reimmunization is supported by best available knowledge of the function of the immune system (Fig. 12) in retaining capability for specific immunologic memory. Two primary doses of vaccine given 1 month apart elicit an initial IgM response that quickly switches to IgG. After a few months, the antibody declines to low level, at which time a third or booster dose of vaccine is given and this results in very rapid antibody response to a very high level. The antibody then falls again to low level or entirely disappears after 5-7 years. At this time, a late booster dose of vaccine or experience with the virus itself quickly sends the antibody soaring to very high levels in a highly effective anamnestic response. Persons exposed to virus at this time show no disease and do not develop a carrier state, though a few may experience very transient infection with antigenemia of short duration plus development of antibody against the core of the virus that is not present in the vaccine. They are protected from clinical disease and from becoming carriers.

Understanding the durability of protection may be facilitated by review of the memory cell system (Fig. 12). Resting B cells that respond to hepatitis B antigen rapidly differentiate to antibody-producing plasma cells upon contact with antigen. The plasma cells are terminally differentiated, have a finite lifespan, and die. The immune system, however, retains a number of partially differentiated cells, called memory cells, that are long-lived and are available for rapid recall and to produce antibody when needed. At each experience with antigen or virus, the memory cell pool is called upon to respond, and, with each cessation, a new bank of memory cells is retained. Thus, immunity is memory! When a late booster dose of vaccine will be needed, if ever, remains to be determined by future experience.

Improving the Immune Response. Considerable attention has been paid to increasing the immune response to hepatitis B vaccine, e. g., in genetic nonresponders, in the immunocompromised such as renal dialysis patients, and in the elderly.

One concept being explored is that of inclusion of polypeptides of the Pre-S1 and Pre-S2 regions (Fig. 3) in the vaccine, with the idea that these might increase the epitope repertoire, especially for T helper epitopes. It is known that the Pre-S2 polypeptide binds to albumin and includes an epitope that induces neutralizing antibody, though of short-term retention. Pre-S1 sequences include the ligand that binds the virus to receptors on liver cells. Various approaches to utilization of Pre-S antigens include the preparation of vaccines that contain whole sequences of selected epitopes from the Pre-S1 and Pre-S2 regions. Recent studies reported at the 1990 International Symposium on Viral Hepatitis and Liver Disease [21] suggest that there may be promise in this approach.

The findings of Meuer et al. [22] are of considerable importance in attempts to understand the failure of immune responses in uremic hemodialysis patients. Meuer and his

Fig. 8. Conceptual structure of hepatitis B surface antigen particle. **Fig. 9.** Function of the immune system (abbreviated). **Fig. 10.** Hypothetical model of MHC class II binding site (From [19]. **Fig. 11.** Dose response in human beings, according to antigen dose and age, given recombinant yeast hepatitis B vaccine

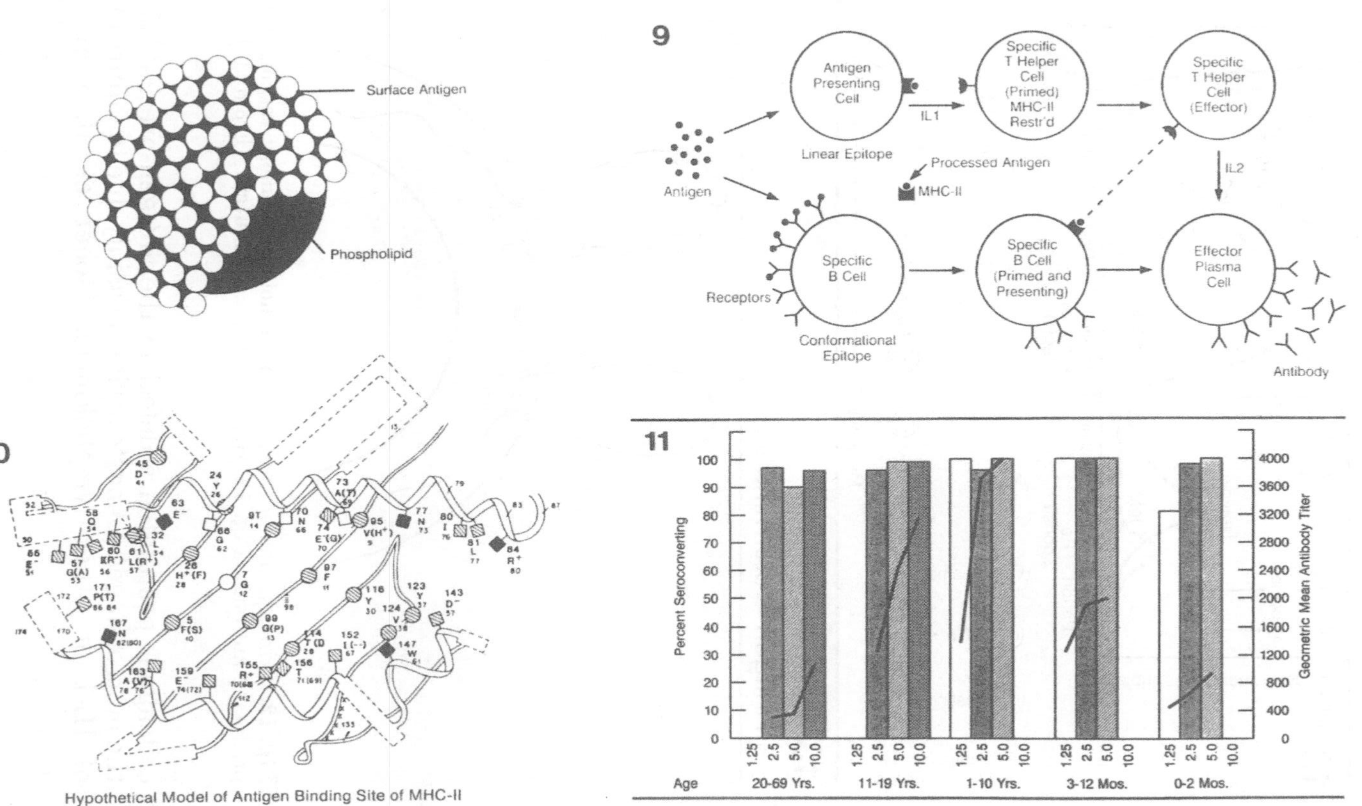

Hypothetical Model of Antigen Binding Site of MHC-II

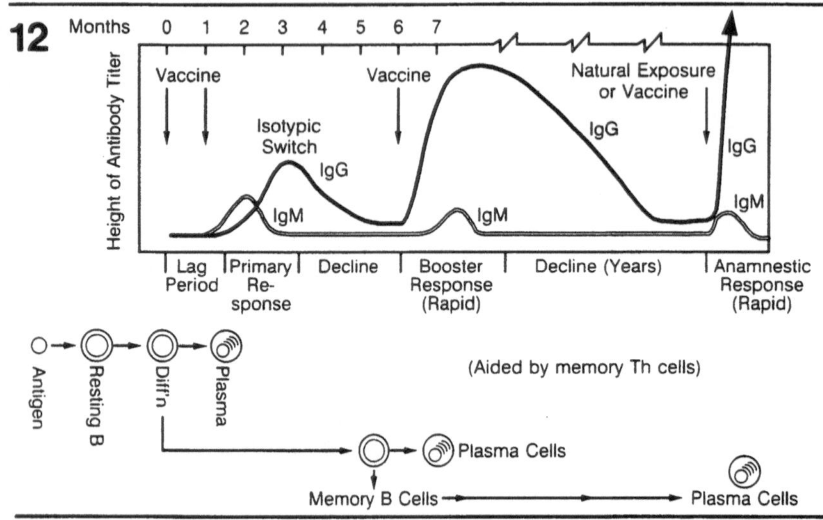

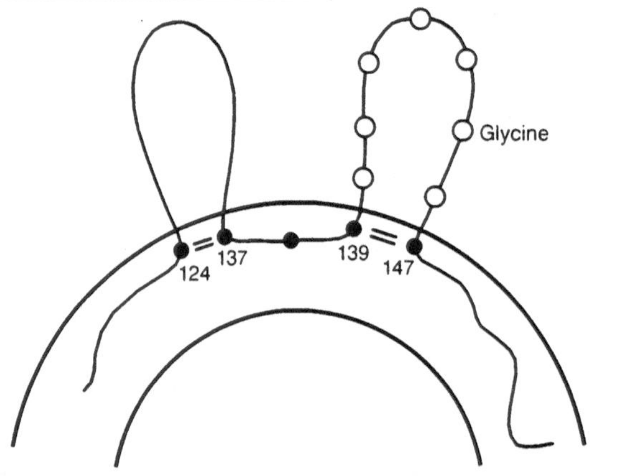

Fig. 12. Primary, anamnestic, and immunologic memory responses of B cells producing antibody. **Fig. 13.** Hepatitis B escape mutant. (From [18])

colleagues postulated a defect in the antigen-presenting cells of the macrophage/monocyte series (possibly due to failure of release of IL-1 or other lymphokines). Meuer et al. were able to circumvent the defect and to obtain normal antibody responses in such immunodeficient patients by immediate second dosing at the injection site with T helper lymphokine, IL-2.

Variant Hepatitis B Serotype. Hepatitis B virus is a double-stranded deoxyribonucleic acid (DNA) virus with the ability to proofread its genetic replication in order to maintain its genetic integrity. It is not unexpected, therefore, that there has been but a single immunologic serotype of hepatitis B virus. It was very surprising, and potentially alarming, however, when Drs. Carman, Zanetti, Zuckerman, Thomas and their coworkers [18] uncovered a second serotype of hepatitis B virus that is not protected against by the hepatitis B vaccine. This variant was first detected in southern Italy and may be moving outside its focal area. Investigation of the basis for immunologic change revealed (Fig. 13) a single-point mutation from guanosine to adenosine at nucleotide position 587, resulting in an amino acid substitution from glycine to arginine in the second *a* antigen loop. The mutation is stable. It is far too early to assess the practical importance of this observation, but it must be borne in mind that a second serotype S antigen might eventually need to be included in hepatitis B vaccine.

Closing Remarks

The events of the past two centuries have resulted in the development of a variety of live attenuated and killed whole virus vaccines. The development of vaccines against hepatitis B has established the rationale and precedents for subunit and recombinant viral vaccines. The initiation of cloning of important segments of the hitherto undescribed viruses of hepatitis C and E [9-11] may point the way to a whole new generation of vaccines against noncultivable and undiscovered agents. There are, however, many important problems and many important viral diseases still to be conquered. For example, can one overcome the inherent dangers from immune enhancement of the disease by dengue virus vaccines? Can the inappropriate and disease-enhancing attributes of respiratory syncytial virus vaccines be overcome by epitope selection? Can the perpetual drift and shift of influenza viruses that render them immunologically obsolete, be overcome, perhaps by focusing on group-reactive cytotoxic T cell responses? Will AIDS virus be overcome in spite of its unique and difficult problems related to viral transmission by infected cells, antigenic hypervariability, viral integration into the host cell genome, selective destruction of the immune system, and replication of the virus in the central nervous system beyond the blood-brain barrier? These are all challenges to the future. But what would the future amount to if there were no challenges?

References

1. de Kruif P (1926) Microbe hunters. Harcourt Brace New York
2. Provost PJ, Hilleman MR (1979) Propagation of human hepatitis A virus in cell culture in vitro. Proc Soc Exp Biol Med 167:213-221
3. Provost PJ, Bishop RP, Gerety RJ, Hilleman MR, Scolnick EM, Stevens CE (1986) New findings in live attenuated hepatitis A vaccine development. J Med Virol 20:165-175
4. Provost PJ, Hilleman MR (1978) An inactivated hepatitis A virus vaccine prepared from infected marmoset liver. Proc Soc Exp Biol Med 159:201-203
5. Hilleman MR, Buynak EB, Roehm RR, Tytell AA, Bertland AU, Lampson GP (1975) Purified and inactivated human hepatitis B vaccine. Progress report. Am J Med Sci 270:401-404
6. Hilleman MR, Bertland AU, Buynak EB, Lampson GP, McAleer WJ, McLean AA, Roehm RR, Tytell AA (1978) Clinical and laboratory studies of hepatitis B surface antigen vaccine. In: Vyas GN, Cohen SN, Schmid R (eds) Viral hepatitis. Franklin Institute, Philadelphia, pp 525-537
7. McAleer WJ, Buynak EB, Maigetter RZ, Wampler DE, Miller WJ, Hilleman MR (1984) Human hepatitis B vaccine from recombinant yeast. Nature 307:178-180
8. Hilleman MR (1987) Yeast recombinant hepatitis B vaccine. Infection 15:3-7
9. Choo Q-L, Kuo G, Weiner AJ, Overby LR, Bradley DW, Houghton M (1989) Isolation of a cDNA clone derived from a blood-borne non-A, non-B viral hepatitis genome. Science 244:359-361
10. Kuo G, Choo Q-L, Alter HJ, Gitnick GL, Redeker AG, Purcell RH, Miyamura T, Dienstag JL, Alter MJ, Stevens CE, Tegtmeier GE, Bonino F, Colombo M, Lee WS, Kuo C, Berger K, Shuster JR, Overby LR, Bradley DW, Houghton M (1989) An assay for circulating antibodies to a major etiologic virus of human non-A, non-B hepatitis. Science 244:362-364
11. Reyes GR, Purdy MA, Kim JP, Luk K-C, Young LM, Fry KE, Bradley DW (1990) Isolation of a cDNA from the virus responsible for enterically transmitted non-A, non-B hepatitis. Science 247:1335-1339

12. Hilleman MR (1986) Vaccinology in practical perspective. Dev Biol Stand 63:5-13
13. Prince AM, Fuji H, Gershon RK (1964) Immunohistochemical studies on the etiology of anicteric hepatitis in Korea. Am J Hyg 79:365-381
14. Prince AM (1968) An antigen detected in the blood during the incubation period of serum hepatitis. Proc Natl Acad Sci USA 60:814-821
15. Blumberg BS, Gerstley BJS, Hungerford DA, London WT, Sutnick AI (1967) A serum antigen (Australia antigen) in Down's syndrome, leukemia and hepatitis. Ann Intern Med 66:924-931
16. Valenzuela P, Medina A, Rutter WJ, Ammerer G, Hall BD (1982) Synthesis and assembly of hepatitis B virus surface antigen particles in yeast. Nature 298:347-350
17. Howard CR, Stirk HJ, Brown SE, Steward MW (1988) Towards the development of synthetic hepatitis B vaccines. In: Zuckermann AJ (ed) Viral hepatitis and liver disease. Liss, New York, pp 1094-1101
18. Carman WF, Zanetti AR, Karayiannis P, Manzillo G, Tanzi E, Zuckerman AJ, Thomas HC (1990) A vaccine-induced escape mutant of hepatitis B virus. Lancet 336:325-329
19. Brown JH, Jardetzky T, Saper MA, Samroui B, Bjorkman PJ, Wiley DC (1988) A hypothetical model of the foreign antigen binding site of class II histocompatibility molecules. Nature 332:845-850
20. West DJ (1989) Clinical experience with hepatitis B vaccines. Am J Infect Control 17:172-180
21. Anonymous (1990) Miscellaneous abstracts. Scientific program and abstract volume. The 1990 international symposium on viral hepatitis and liver disease, 4-8 April, 1990. Houston pp 116-119
22. Meuer SC, Dumann H, Meyer zum Buschenfelde KH, Kohler H (1989) Low dose interleukin-2 induces systemic immune responses against HBsAg in immunodeficient non-responders to hepatitis B vaccine. Lancet 1:15-17

Immunopathogenesis of HIV/SIV Infections: Consequences for the Development of AIDS Vaccines

R. Kurth, J. Ennen, J. Denner, S. Hartung, D. Binninger, M. Vogel and S. Norley

Paul-Ehrlich-Institut, Paul-Ehrlich-Str. 51-59, 6070 Langen, FRG

Introduction

The pathogenetic mechanisms by which human immunodeficiency virus (HIV) finally impairs the immune response are not yet understood. Besides the direct killing of infected CD4+ T-lymphocytes, several indirect mechanisms for the functional impairment of the immune system have by now been recognized. The outer envelope glycoprotein of HIV can easily dissociate from the virus particle and may subsequently bind to the CD4 receptor of lymphocytes, thus rendering the uninfected cell vulnerable to immune attack [1]. Furthermore, the infection of follicular dendritic reticulum cells and of Langerhans cells may well impair lymphocyte maturation and antigen recognition, respectively [2]. A recently demonstrated HIV infection of lymphocytic bone marrow precursor cells may also lead to inadequate lymphocyte differentiation and maturation, which in turn may result in an incomplete supply of new CD4+ T-lymphocytes [3].

Therefore, it was fortunate that simian immunodeficiency viruses (SIV) were detected in two natural hosts. SIV_{mn} was isolated from mandrills and SIV_{agm} from African green monkeys (AGMs). These monkeys are infected in the wild and yet remain healthy. Our research has concentrated on SIV_{agm} in AGMs with the aim of understanding why these animals remain healthy despite prolonged lentivirus infection. Learning why these monkeys remain healthy may increase our understanding of why HIV in humans is able to induce immune suppression and disease. Several working hypotheses spring to mind which are experimentally testable and which may help to explain the basis for the lack of SIV_{agm} pathogenicity. First, it is conceivable that AGMs launch a particularly vigorous immune response. Second, the host cell tropism of SIV_{agm} may be different from HIV in humans. Third, the in vivo variability of SIV_{agm} may be reduced compared with HIV, preventing the development of virus escape mutants. Fourth, the now well-recognized immunosuppressive domain of the HIV transmembrane protein may be absent in SIV_{agm}. Finally, it is conceivable that AGMs are better equipped to suppress intracellularly virus replication, thus reducing in vivo viral load.

The Immune Response to HIV and SIV_{agm}

There is sufficient clinical evidence to suggest that in humans the immune response is at least initially able to limit HIV replication. Indeed, the immune system may be responsible for the long latency period between infection and disease manifestation. There are long-term infected individuals without disease development, and even in persons with acquired immunodeficiency syndrome (AIDS)-related complex, stable clinical situations can be observed. Although other mechanisms of viral repression may be possible, low peripheral virus load during the latent phase [4] and the development of virus variants in vivo [5-8] are indicative of immune surveillance. What we still need to know, however, is which of the different immune response mechanisms may eliminate free HIV particles and which HIV-infected cells. It is also not yet known which HIV antigens induce a protective immune response and how HIV eventually escapes the immune surveillance.

Neutralizing antibodies, complement-activating cytotoxic antibodies, antibody-dependent cellular cytotoxicity (ADCC), natural killer cell (NK) activity and cytotoxic T-lymphocytes (CTLs) represent the possible immunological anti-viral effector mechanisms. We have studied these effector mechanisms in humans and in AGMs in some detail. Binding/neutralizing antibodies in HIV-infected humans are directed only against a few epitopes, predominantly in the outer envelope glycoprotein gp120 [9-11]. Approximately 80% of the infected individuals develop neutralizing antibodies which broaden in specificity to react with more and more HIV variants over time. However, the neutralizing antibody titer is relatively low compared with other virus infections. There is no clear correlation between antibody titer or specificity and disease progression [12]. Some preliminary evidence indicates, however, that antibodies specific for a certain region of the gp120 (epitopes in the hypervariable V3 loop of gp120) may prevent intrauterine and/or perinatal HIV transmission [13]. Except for three select cases [14], neutralizing antibodies have not been shown to protect immunized chimpanzees against subsequent virus challenge.

It was surprising to learn that in AGMs neutralizing antibodies are even less frequent and of a lower titer than in HIV-infected humans [15]. In only about 10% of the infected animals do neutralizing antibodies develop with titers up to 1:40. It is therefore likely that in AGMs neutralizing antibodies do not contribute to the limitation of virus spread or to prevention of disease development.

In contrast to the situation found in animals infected with the oncovirus subgroup of retroviruses, the leukemia-sarcoma viruses, cytotoxic complement-activating antibodies develop neither in HIV-infected humans [16] nor in SIV_{agm}-infected AGMs [15]. In other systems these antibodies are able to initiate lysis of infected cells by the complement cascade and it would be desirable to induce such an immune response by appropriate immunizations (see below).

ADCC is well-developed in infected humans and AGMs [15]. The antibodies recognize viral antigens on the infected cell surface and, in conjunction with effector cells, eliminate virus-infected lymphocytes. ADCC activity has been found to be present in all sera from HIV- and SIV_{agm}-infected individuals [12] and the antibody titers are relatively high (up to $1:10^6$). The reactivity of the corresponding antibody populations is relatively broad, i.e., specific for all HIV-infected cells. Similarly, ADCC activity in infected AGMs will kill all infected cells, regardless of which SIV_{agm} strain has been used for target cell infection. ADCC can be induced in chimpanzees with purified gp120 *env* [16]. Despite these encouraging activities, there is, unfortunately, no correlation between ADCC activity and enzyme-linked immunosorbent assay (ELISA) or neutralizing antibody titers. More important, a correlation between clinical status and ADCC activity could not be established [12]. Interestingly, in a study of a cohort of HIV-infected men, about 5% exhibited an ADCC activity that was able to destroy not only HIV-1 infected target cells, but also HIV-2 infected cells [12]. It remains to be seen whether those few people have a better clinical prognosis than others with an ADCC activity restricted to HIV-1 infected target cells only.

NKs are also present in the majority of HIV-infected humans. This activity eliminates HIV-infected cells and can further be stimulated by the addition of monokines and lymphokines to corresponding in vitro assays. The effector cells belong to the CD16+ lymphocyte subpopulation and the activity declines in later stages of disease progression [17]. The specificity of the NK activity is likely to be directed against epitopes of the gp120 *env*, as the addition of purified gp120 *env* inhibits lysis by NK cells.

CTLs represent a potentially very effective antiviral defense mechanism. After HIV infection, CTLs develop that recognize epitopes in the structural envelope and *gag* proteins [18-20] in the *pol* gene product reverse transcriptase [21], and in the regulatory viral proteins of *nef* and *vif* [22-23]. Similarly, CTLs specific for the SIV_{agm} envelope protein have been demonstrated (Norley et al., unpublished data). The exact specificity of CTLs remains to be elucidated, but it can be expected that, in contrast to neutralizing antibodies, the CTLs may be more broadly reactive, as some of the recognized viral epitopes are located in relatively conserved amino acid domains of HIV and SIV, for example in *gag*. The effector cells are predominantly of the CD8+/CD4-phenotype, although CD8-/CD4+ CTL subpopulations have also been isolated.

CTLs are particularly effective in combatting viral infections, eliminating virus-infected cells via viral protein fragments expressed on the cell surface. In vitro, CTLs have been shown to interrupt HIV transmission by cell-to-cell contacts [24] and CD8+ CTL activity could be correlated with suppression of virus replication in vitro and in vivo. After 3'-azido-3'-deoxythymidine (AZT) treatment an increased CTL activity was shown to parallel prolonged survival and a decreased CTL activity was associated with development of AIDS and simian AIDS (SAIDS). On the negative side, neutralizing antibodies, ADCC and CTL activities have been shown to participate in "innocent bystander" killing of noninfected CD4+ lymphocytes that have absorbed, processed, and re-expressed viral antigens. In other words, these immune reactivities are also directed against uninfected lymphocytes.

In summary, there are no major qualitative or quantitative differences in the immune responses against HIV and SIV_{agm} in humans and AGMs, respectively. We therefore assume at present that it is not a vigorous immune response in AGMs that prevents disease development after SIV_{agm} infection.

The Immunosuppressive Peptide of Retroviruses

The transmembrane proteins of retroviruses (p15E and gp41/45) contain an evolutionarily conserved amino acid sequence which, when tested as synthetic peptide, exerts immunosuppressive effects in in vitro assays [25]. The immunosuppressive domain of HIV is located in the amino acids 583-599 of the HIV *env* gene product. The peptide inhibits mitogen and antigen stimulation of human lymphocytes [26,27] and is also interspecies-active, i.e., it inhibits mouse lymphocyte function in vitro (Fig. 1). Due to the evolutionary conservation, we were able to localize a corresponding amino acid sequence in SIV_{agm}. The still preliminary in vitro assays indicate that the immunosuppressive peptide of SIV_{agm} is indeed much less immunosuppressive in vitro than the corresponding peptide of HIV. Similar differences in the in vitro immunosuppressive activity have been shown for inactivated noninfectious virus preparations of HIV-1 and SIV_{agm} [15]. It certainly remains to be seen (e.g., by recombinant HIV/SIV_{agm} virus hybrids) whether the transmembrane proteins of the lentiviruses contribute to immunosuppression in vivo.

Host Cell Tropism of HIV and SIV_{agm}

A detailed analysis of the tropism of HIV and SIV_{agm} has not yet revealed significant differences. Both subgroups of lentiviruses infect predominantly monocytes and macrophages and the CD4+ T-lymphocyte subset [28,29]. In addition, HIV-1 infects

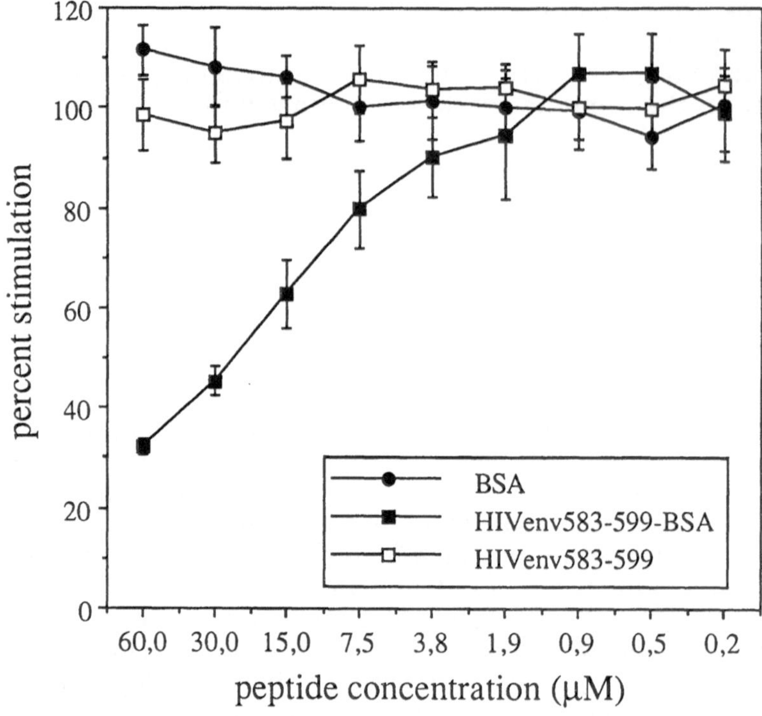

Fig. 1. Inhibition of human lymphocyte concanavalin A stimulation by the isopeptide HIV *env* 583-599 conjugated to BSA. Unconjugated soluble peptide and carrier BSA alone are not immunosuppressive

Langerhans and follicular dendritic reticulum cells [2]. It is as yet unknown whether SIV_{agm} can also infect microglial cells as can HIV. In any case, according to our present knowledge, a difference in tropism cannot be held responsible for the different pathogenic potential of HIV and SIV_{agm}. However, one (perhaps important) difference between the interactions of HIV-1 and SIV_{agm} with their host cells is that, whereas incubation of soluble CD4 with HIV-1 prior to exposure to cells blocks any subsequent infection (as would be expected), SIV_{agm} infectivity is *enhanced* by the presence of soluble CD4 [30].

Table 1. Titers of infectious SIV_{agm} in peripheral blood mononuclear cells and plasma of 17 seropositive African green monkeys

Animal no.	PBMC titer ($TCID/10^6$ cells)	Plasma titer (TCID/ml)
8	0	0
15	0	0
33	0	0
2a	1	20
4	1	0
26	1	0
Z2	1	0
Z8	10	0
16	10	0
24	10	0
43	10	2
Z4	10	0
Z13	10	0
Z14	10	0
3	20	0
37	20	2
38	20	0.2

PBMC, peripheral blood mononuclear cells; TCID, tissue culture infective dose

Development of In Vivo Variability

We were able to construct a molecular SIV_{agm} clone that retained infectivity in vitro and in vivo. Using this clone, animals were infected and the development of in vivo variability was investigated. We repeatedly and directly sequenced viral DNA from these animals and found that point mutations accumulated in the animals over time. Variability from the single injected viral genotype developed with a maximal rate of 7.7 mutations per 1 000 nucleotides per year [31]. A remarkable selection for nonsynonymous mutations, which accounted for 92% of all changes, indicated continuous selection of new variants. Due to the absence of a particularly overwhelming immune pressure, and to the absence of any correlation between immunodominant epitopes and variable domains [32], it is possible that the selection pressure is non-immunological in nature.

Intracellular Virus Suppression

Intracellular control of virus replication by still virtually unknown mechanisms may lead to reduced in vivo viral load and thus to an inhibition of the pathogenic consequences of virus infection. We therefore continued our comparison of HIV and SIV_{agm} infections by quantitating the peripheral virus loads. The SIV_{agm} load in peripheral blood mononuclear cells and in plasma (see Table 1) is a little lower (though in the same range) as that seen in asymptomatic HIV-1 patients and much lower than in patients with AIDS-related complex (ARC) and AIDS [33]. SIV_{agm} infected monkeys are therefore in a similar situation as healthy HIV-1 infected humans. Why HIV-1 should at some stage switch to the rapid replication and high viral load associated with disease and SIV_{agm} not is as yet unknown.

We have initiated a study of the proviral DNA in freshly HIV/SIV_{agm} infected peripheral blood monocytic cells and in lymphocytic cell lines with the aim of elucidating potential intracellular repressor mechanisms. It is obvious that the identification of cellular protein suppressing virus replication may help to explain viral pathogenesis and may even have potential therapeutic value.

Fig. 2. Attenuation of SIV_{mac} by *nef* deletion. Cloned mutant viral DNA is transfected into a suitable plasmid-expressing recipient cell. Progeny mutant virus is produced that upon infection of SIV-susceptible target cells (e.g., lymphocytes, macrophages) should immediately continue to replicate without being able to undergo latency. **Fig. 3.** Attenuation of SIV_{mac} by *int* deletion and complementation. Mutant viral DNA is transfected into a recipient cell which has previously been transfected with an integrase-expressing plasmid. Thus, phenotypically mixed (*int* protein containing) virus particles will be synthesized. They will be competent to infect and integrate their proviral DNA in the next round of infection. These cells serve as stable producers of (large amounts) of SIV Δ*int* particles which upon infection of tissue cultures or animals will only transiently replicate from episomal SIV DNA, limiting in vivo infection

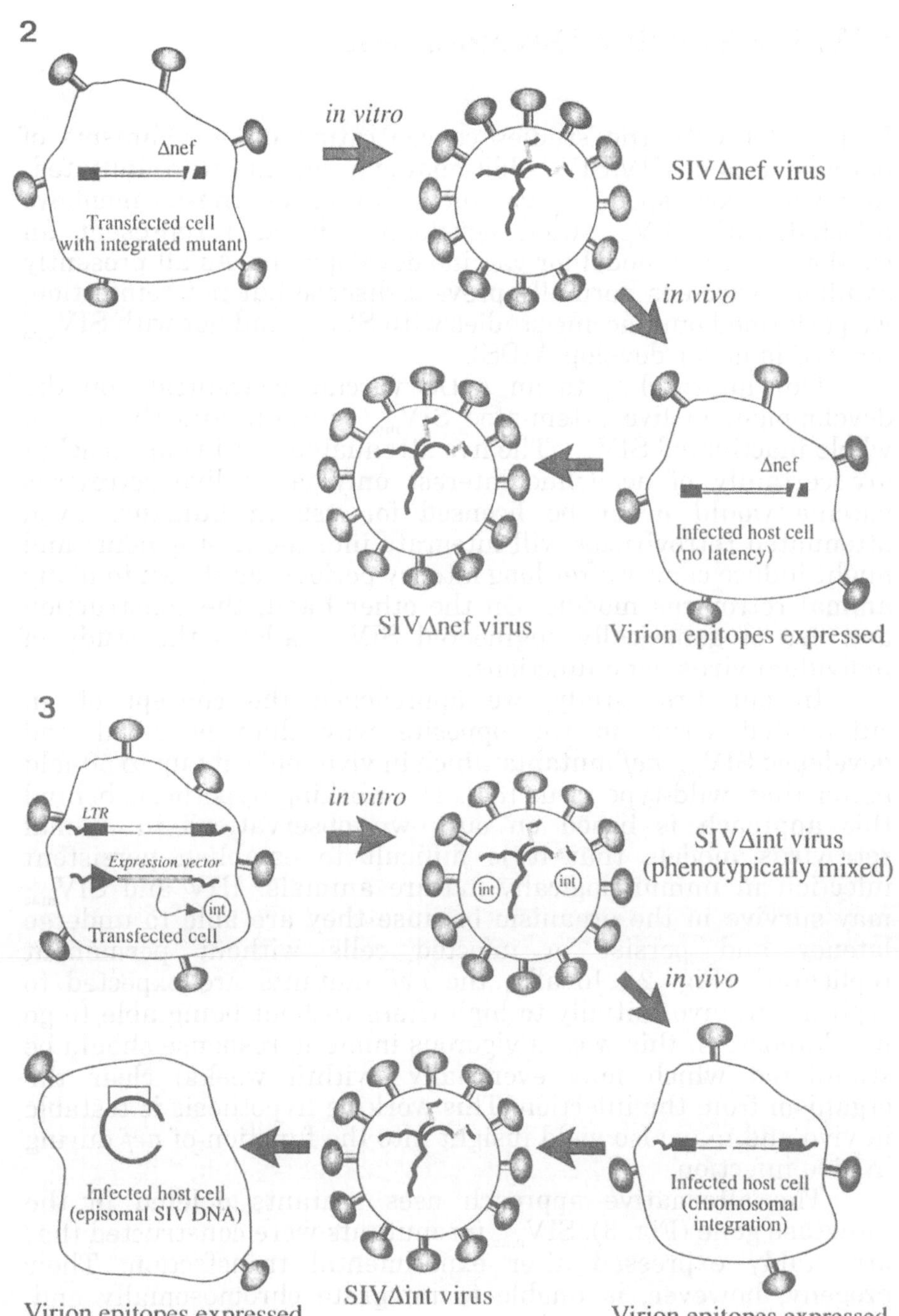
2
Δnef
Transfected cell
with integrated mutant
in vitro
SIVΔnef virus
in vivo
Δnef
Infected host cell
(no latency)
SIVΔnef virus
Virion epitopes expressed
3
LTR
Expression vector
int
Transfected cell
in vitro
int
int
SIVΔint virus
(phenotypically mixed)
in vivo
Infected host cell
(chromosomal
integration)
Virion epitopes expressed
SIVΔint virus
Infected host cell
(episomal SIV DNA)
Virion epitopes expressed

HIV/SIV Vaccine Development

In parallel with the studies concentrating on mechanisms of pathogenesis of HIV/SIV, additional experiments were initiated to develop corresponding vaccines. Nowadays, rhesus monkeys infected with SIV_{mac} (and developing disease) represent an excellent animal model for vaccine development. As all presently available vaccines normally prevent disease but not reinfection, we performed our vaccine studies with SIV_{mac} and not with SIV_{agm} (as AGMs never develop AIDS).

Our approaches to an AIDS vaccine concentrate on the development of live attenuated SIV_{mac} mutants and the use of whole inactivated SIV_{mac}. The live attenuated vaccine approaches are certainly of academic interest only, as a live retrovirus vaccine would never be licensed for use in humans. Even attenuated retroviruses will integrate into the host genome and might induce cancer after long latency periods, as shown in many animal retrovirus models. On the other hand, the construction and use of genetically engineered SIV_{mac} allows the study of individual virus gene functions.

In our first study, we approached the concept of an 'attenuated virus' in the opposite way than is usual and developed SIV_{mac} *nef* mutants which in vivo replicate up to 50-fold better than wild-type virus [34]. The working hypothesis behind this approach is based on our own observations in animal retrovirus models that it is difficult to establish persistent infection in immunologically mature animals. HIV and SIV_{mac} may survive in the organism because they are able to undergo latency and persist in infected cells without permanent replication (Fig. 2). Ideally, the *nef* mutants are expected to replicate in vivo initially to high titers without being able to go into latency. In this way, a vigorous immune response should be stimulated which may eventually (within weeks) clear the organism from the infection. This working hypothesis is testable in vivo and may also yield insight into the function of *nef* during in vivo infection.

The alternative approach uses mutants deleted in the integrase gene (Fig. 3). SIV_{mac} *int* mutants were constructed that are stably expressed after experimental transfection. Their progeny, however, is unable to integrate chromosomally and, according to our preliminary data, continues to replicate for only a few weeks, thus limiting infection by the attenuated 'vaccine virus'. Starting with transfected cells that complement *int* mutants with *int* proteins, we can now use their virus-containing supernatant to infect animals either directly or their lymphocytes first in vitro, before giving the cells to the animals. The corresponding immune response can subsequently be analysed, and animals with a well-developed immune response will be subjected to wild-type virus challenge.

In a more down-to-earth approach, we, like others, used whole inactivated SIV_{mac} to immunize rhesus monkeys. In a first series of experiments, we used muramyl dipeptide as adjuvant and the animals were immunized with two different doses of SIV_{mac} (for details see Table 2). After three immunizations, animals were challenged with 33 infectious particles by the intravenous route attempts, by analyzing the animals' immunologic (memory) response and by direct polymerase chain reaction for viral DNA.

It was encouraging to see that all four animals immunized with the higher SIV_{mac} dose were apparently fully protected against subsequent infection [35]. The detailed humoral and cellular immune response of the protected and non-protected animals is presently being investigated. At first sight, however, neutralizing antibodies do not appear to be (solely) responsible for the protection observed.

Obviously, many questions remain to be answered for the development of a SIV/HIV vaccine. Our group will next concentrate on using better adjuvants and on trying to protect against challenge by relatively heterogeneous virus variants and against infected lymphocytes. There are still many difficulties that need to be overcome for the successful development of an HIV vaccine (for details see [36] and [37]). Nevertheless, the successful demonstration that animals can be immunized against subsequent lentivirus infection warrants further studies and increased efforts to develop our capacity to limit the otherwise relatively uninhibited spread of HIV by immunoprophylactic means.

Table 2. Clinical and immunological characteristics of rhesus monkeys vaccinated with different doses of inactivated SIV_{mac}

Rhesus monkey	Immunogen (mg)	Homologous neutralising antibody titer	Lymphadenopathy after challenge	Virus reisolation	Anamnestic response	Protection against challenge
Rh25	0.1	269	yes	yes	yes	no
Rh27	0.1	226	yes	yes	yes	no
Rh36	0.1	57	yes	yes	yes	no
Rh37	0.1	190	no	no	no	yes
Rh28	0.5	761	no	no	no	yes
Rh30	0.5	269	no	no	no	yes
Rh34	0.5	453	no	no	no	yes
Rh35	0.5	190	no	no	no	yes
Rh24	none	0	yes	yes	n.a.	no
Rh33	none	0	yes	yes	n.a.	no

[a]Inactivated SIV_{mac} in muramyl dipeptide inoculated at 0, 4, 8, and 16 weeks

References

1. Lyerly HK, Matthews TJ, Langlois AJ, Bolognesi DP, Weinhold KJ (1987) Human T-cell lymphotropic virus IIIB glycoprotein (gp120) bound to CD4 determinants on normal lymphocytes and expressed by infected cells serves as target for immune attack. Proc Natl Acad Sci USA 84:4601-4605
2. Macatonia SE, Lau R, Patterson S, Pinching AJ, Knight SC (1990) Dendritic cell infection, depletion and dysfunction in HIV-infected individuals. Immunology 71:38-45
3. Schnittman SM, Denning SM, Greenhouse JJ, Justement JS, Baseler M, Kurtzberg J, Haynes BF, Fauci AS (1990) Evidence for susceptibility of intrathymic T-cell precursors and their progeny carrying T-cell antigen receptor phenotypes TCR alpha beta + and TCR gamma delta + to human immunodeficiency virus infection: a mechanism for CD4+ (T4) lymphocyte depletion. Proc Natl Acad Sci USA 87:7727-7731
4. Brinchmann JE, Albert J, Vartdal F (1991) Few infected CD4+ T cells but a high proportion of replication-competent provirus copies in asymptomatic human immunodeficiency virus type 1 infection. J Virol 65:2019-2023
5. Albert J, Abrahamsson B, Nagy K, Aurelius E, Gaines H, Nystroem G, Fenyoe EM (1990) Rapid development of isolate-specific neutralizing antibodies after primary HIV-1 infection and consequent emergence of virus variants which resist neutralization by autologous sera. AIDS 4:107-112
6. Pircher H, Moskophidis D, Rohrer U, Buerki K, Hengartner H, Zinkernagel RM (1990) Viral escape by selection of cytotoxic T cell-resistant virus in vivo. Nature 346:629-633
7. Saag MS, Hahn BH, Gibbons J, Li YX, Parks ES, Parks WP, Shaw GM (1988) Extensive variation of human immunodeficiency virus type-1 in vivo. Nature 334:440-444
8. Wain-Hobson S (1989) HIV genome variability in vivo. AIDS 3 [Suppl]
9. Ho DD, Kaplan JC, Rackauskas IE, Gurney ME (1988) Second conserved domain of gp120 is important for HIV infectivity and antibody neutralization. Science 239:1021-1023
10. Ho DD, Sarngadharan MG, Hirsch MS, Schooley RT, Rota TR, Kennedy RC, Chanh TC, Sato VL (1987) Human immunodeficiency virus neutralizing antibodies recognize several conserved domains on the envelope glycoproteins. J Virol 61:2024-2028
11. Rusche JR, Javaherian K, McDanal C, Petro J, Lynn DL, Grimaila R, Langlois A, Gallo RC, Arthur LO, Fischinger PJ, et al (1988) Antibodies that inhibit fusion of human immunodeficiency virus-infected cells bind a 24-amino acid sequence of the viral envelope, gp120. Proc Natl Acad Sci USA 85:3198-3202
12. Norley SG, Mikschy U, Werner A, Staszewski S, Helm EB, Kurth R (1990) Demonstration of cross-reactive antibodies able to elicit lysis of both HIV-1- and HIV-2-infected cells. J Immunol 145:1700-1705
13. Rossi P, Moschese V, Broliden PA, Fundaro C, Quinti I, Plebani A, Giaquinto C, Tovo PA, Ljunggren K, Rosen J, et al (1989) Presence of maternal antibodies to human immunodeficiency virus 1 envelope glycoprotein gp120 epitopes correlates with the uninfected status of children born to seropositive mothers. Proc Natl Acad Sci USA 86:8055-8058

14. Emini EA, Schleif WA, Murthy K, Eda Y, Tokioshi S, Putney SD, Matsushita S, Nunberg JH, Eichberg JW (1991) Passive immunisation with a monoclonal antibody directed to the HIV-1 gp120 principal neutralisation determinant confers protection against HIV-1 challenge in chimpanzees. VIIth International AIDS Conference, Florence, Italy (Abstract TH.A.64.)
15. Norley SG, Kraus G, Ennen J, Bonilla J, König H, Kurth R (1990) Immunological studies of the basis for the apathogenicity of simian immunodeficiency virus from African green monkeys. Proc Natl Acad Sci USA 87:9067-9071
16. Nara PL, Robey WG, Gonda MA, Carter SG, Fischinger PJ (1987) Absence of cytotoxic antibody to human immunodeficiency virus-infected cells in humans and its induction in animals after infection or immunization with purified envelope glycoprotein gp120. Proc Natl Acad Sci USA 84:3797-3801
17. Sirianni MC, Soddu S, Malorni W, Arancia G, Aiuti F (1988) Mechanism of defective natural killer cell activity in patients with AIDS is associated with defective distribution of tubulin. J Immunol 140:2565-2568
18. Clerici M, Lucey DR, Zajac RA, Boswell RN, Gebel HM, Takahashi H, Berzofsky JA, Shearer GM (1991) Detection of cytotoxic T lymphocytes specific for synthetic peptides of gp160 in HIV-seropositive individuals. J Immunol 146:2214-2219
19. Miller MD, Lord CI, Stallard V, Mazzara GP, Letvin NL (1990) The *gag*-specific cytotoxic T lymphocytes in rhesus monkeys infected with the simian immunodeficiency virus of macaques. J Immunol 144:122-128
20. Nixon DF, Townsend AR, Elvin JG, Rizza CR, Gallwey J, McMichael AJ (1988) HIV-1 *gag*-specific cytotoxic T lymphocytes defined with recombinant vaccinia virus and synthetic peptides. Nature 336:484-487
21. Hosmalin A, Clerici M, Houghten R, Pendleton CD, Flexner C, Lucey DR, Moss B, Germain RN, Shearer GM, Berzofsky JA (1990) An epitope in human immunodeficiency virus 1 reverse transcriptase recognized by both mouse and human cytotoxic T lymphocytes. Proc Natl Acad Sci USA 87:2344-2348
22. Culmann B, Gomard E, Kieny MP, Guy B, Dreyfus F, Saimot AG, Sereni D, Levy JP (1989) An antigenic peptide of the HIV-1 NEF protein recognized by cytotoxic T lymphocytes of seropositive individuals in association with different HLA-B molecules. Eur J Immunol 19:2383-2386
23. Culmann B, Gomard E, Kieny MP, Guy B, Dreyfus F, Saimot AG, Sereni D, Sicard D, Levy JP (1991) Six epitopes reacting with human cytotoxic CD8+ T cells in the central region of the HIV-1 NEF protein. J Immunol 146:1560-1565
24. Walker BD, Plata F (1990) Cytotoxic T lymphocytes against HIV. AIDS 4:177-184
25. Denner J (1987) Immunosuppression by oncogenic retroviridae. In: Zschiesche et al. (eds.) Immune modulation by infectious agents. Fischer, Jena, pp 140-201
26. Ruegg CL, Monell CR, Strand M (1989) Inhibition of lymphoproliferation by a synthetic peptide with sequence identity to gp41 of human immunodeficiency virus type 1. J Virol 63:3257-3260
27. Denner J, Kurth R (1990) Suppression of IL-2 dependent proliferation of mouse T-lymphocytes by a HIV-1 gp41-derived synthetic peptide. VIII International Virology Congress, Berlin, FRG (Abstract W58-006)

28. Kraus G, Werner A, Baier M, Binninger D, Ferdinand FJ, Norley S, Kurth R (1989) Isolation of human immunodeficiency virus-related simian immunodeficiency viruses from African green monkeys. Proc Natl Acad Sci USA 86:2892-2896
29. Werner A, Winskowsky G, Cichutek K, Norley SG, Kurth R (1990) Productive infection of both CD4+ and CD4- human cell lines with HIV-1, HIV-2 and SIVagm. AIDS 4:537-544
30. Werner A, Winskowsky G, Kurth R (1990) Soluble CD4 enhances simian immunodeficiency virus SIV_{agm} infection. J Virol 64:6252-6256
31. Baier M, Dittmar MT, Cichutek K, Kurth R (1991) Development in vivo of genetic variability of simian immunodeficiency virus. Proc Natl Acad Sci USA 88:8126-8130
32. Denner J, Baier M, Dittmar M, Cichutek K, Norley S, Kurth R (1991) Genetic variability of molecular SIVagm clones in vivo. VIIth International AIDS Conference, Florence (Abstract TH.A.28)
33. Hartung S, Boller K, Cichutek K, Norley S, Kurth R (1991) Quantitation of a lentivirus in its natural host: SIV_{agm} in African green monkeys. J Virol 66:2143-2149
34. Binninger D, Ennen J, Bonn D, Norley S, Kurth R (1991) Mutational analysis of the SIV_{mac} *nef* gene. J Virol 65:5237-5243
35. Hartung S, Norley SG, Ennen J, Cichutek K, Plesker R, Kurth R (1991) Vaccine protection against SIV_{mac} infection by high but not low dose whole inactivated virus immunogen. J AIDS (in press)
36. Kurth R, Binninger D, Ennen J, Denner J, Hartung S, Norley S (1991) The quest for an AIDS vaccine: the state of the art and current challenges. AIDS Res Hum Retroviruses 7:425-433
37. Norley S, Kurth R (1991) Neutralizing antibodies and antigens in AIDS. Infection 19:583-588

Subject Index